DOUBLE MENACE

P-82 Twin Mustang

by David R. McLaren

ISBN 0-934575-12-6

ViP Publishers, Inc.
P.O. Box 16103
Colorado Springs, Colorado 80935

Table of Contents

To Matilda Rose McLaren

Acknowledgements

Researching an aircraft as relatively obscure as the Twin Mustang became quite a challenge and would have proven impossible without the help of the following former employees of North American Aviation: Norm Avery, Gene Boswell, Bob Chilton, Ed Horkey, Dick Schleicher and Ed Virgin.

Additional support came from the following aviation photographers and historians. Gerald T. Cantwell (Air Force Reserve), Kathy Cassity (Air Force Museum), Paul Coggan (Mustang International), Roger Besecker, Don Brackenreg, B.J. Buckhout, Bob Burns (NASA), Larry Davis, John de Vries, Bob Esposito, Carl Frazer, Bill Greenhalgh, Marty Isham, Bill Larkins, Robert Loffredo, Grant Hales, David W. Menard, Merle Olmstead, B.C. Reed, Harry White, Roy Wubker and Joe Quartuccio.

Introduction

There is a saying within the aviation community that if an airplane looks right, it will fly right. The P-82 Twin Mustang became somewhat of an exception to this cliché, for when viewed from some angles it was beautiful, while from others it appeared quite odd, to say the least. It was an unusual design, particularly for an aircraft which saw mass production, for most of the more unusual configurations were strictly "one off" by the manufacturer for various research or development projects.

When everything was operating correctly, the F-82 flew like a dream, but if it was out of tune, either through engine or airframe problems, it earned the nickname "The Beast."

Many Aviation Buffs have trouble envisioning this fighter, for little is known of its history. In some instances it is confused with the Lockheed P-38 or the Northrop P-61, although it was far superior to either of these aircraft. Also, there is a certain resemblance in the fact that all three aircraft were twin-boomed and twin-engined, and there were all-weather versions of each example.

The Twin Mustang was 100 miles per hour faster than the P-38 in the day fighter role, and the all-weather version of the F-82 had a good 150 mile an hour advantage over the P-61, the aircraft it eventually replaced in that function.

From the standpoint of range, in both miles to be covered and the time it could remain in the air, the P-82 could exceed the totals of the other two aircraft combined. In fact, the World Speed record between Hawaii and New York City, established by a F-82B has never been broken by any other conventionally powered aircraft.

As capabilities go, the F-82 was the ultimate piston-powered fighter, and the last one ordered by the United States Air Force. It could fly higher, faster and further than any other conventional fighter, and the F-82E was the only fighter capable of intercepting the intercontinental B-36 on the B-36's terms, and then make repeated attacks upon the Peacemaker until the exercise was terminated. (This was, of course, true only until the F-86 became operational, at which time a new era of aerial warfare began.) In today's era of F-14's through F-22's such a statement may not appear all that impressive, but the F-82 could outperform the USAF's early F-80's and F-84's in total, along with anything the US Navy or any other country had to offer. This was much to the embarrassment of both the jet advocates and the USN during the great "Battleship Controversy" of 1948.

But within two years of this time, the F-82's zenith had been passed and it was beginning to be phased out in favor of improved jet powered aircraft. All that was left for it was to help hold the line in the Korean conflict until enough jet fighters were available to take its place, and it did that very well indeed.

F-82 Unit Insignia

F-82 Unit Insignia

Definition - The P/F-82 Twin Mustang - a twin engined, low wing fighter, distinguished by its two fuselages, each with its own cockpit. It was operational in the USAF from the late 40's through the mid-50's in a variety of roles. Although many versions of the aircraft had dual controls (and some had an autopilot) it was usually flown from the left hand fuselage. With its four wheeled landing gear (two mains, two tailwheels) there was a tendency for the pilot to land "his side" of the airplane first!

Program Inception

The P-82, which became known as the F-82 during the USAF reclassification program of "Pursuit" to "Fighter" in 1947, had its genesis in 1943 when the United States Army Air Force submitted a request to the aviation industry for a long-range fighter for the Pacific Theater of Operations. It was hoped that General Douglas MacArthur's proposed island-hopping campaign would be successful enough to place B-29 Superfortress' close enough to Japan to carry out extensive bombing raids over the enemy islands. However it had already been proven in Europe that bombers could not operate without an escort – and a fighter with longer legs than either the P-47 or P-51 would be required.

as the RAF had been forced to do in Europe. This continued until Iwo Jima was captured in March, 1945 and fighter escorts, in the form of P-51D's, became available. Even so, these were long over-water missions for the P-51's and many of the targeted areas in Japan were beyond their range. What was really needed was a fighter that had the range, speed, maneuverability and general ability to shepherd the bombers for whatever distance was required. This need was particularly emphasized in light of the impending invasion of Japan itself, scheduled to commence with Operation Olympic on November 1, 1945.

North American Aviation submitted their design

The Twin Mustang

As it was, B-29's started combat operations out of China and the Marianas Islands against Japan in November, 1944 without a great deal of success. This was partially due to the lack of a fighter escort. The USAAF had to revert to night bombing, much

to General "Hap" Arnold on January 7, 1944. Representing NAA was James H. "Dutch" Kindelberger, North American's President, and Ed Schmued, Chief of Design Engineering, to whom credit for the Twin Mustang belongs. To say that it was an unusual proposal would be an understatement, for although the idea of joining two aircraft fuselages on one wing had been accomplished by the Luftwaffe with their five-engined Heinkel He 111 Z-1 and their

thodox lines. The effort, as odd as it was, was sold to Arnold for several reasons. Development time would be reduced through having the knowledge already gained through NAA's P-51 program. It was pointed

P-82 Mock-up

twin-engined 1942 Messerschmitt Me 109 Z, North American Aviation was the first and only U.S aircraft company to submit a design along such unorthodox lines.

out that if one P-51 was good, then two would be even better. In addition, retooling would be kept to a minimum, as would be the design of component

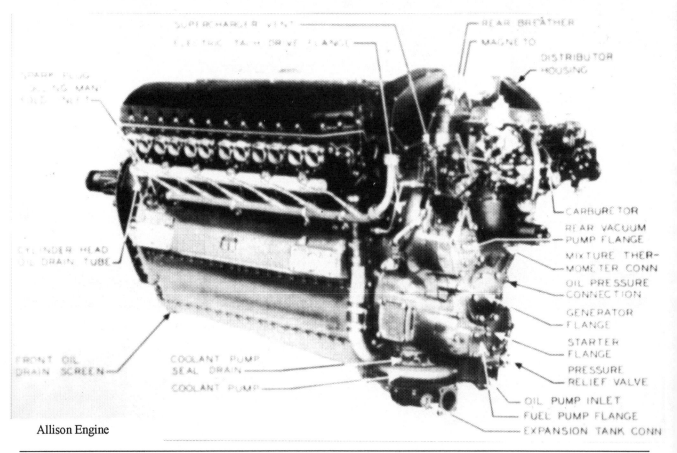

Allison Engine

parts, such as propellers, engines and fuel cells. Another plus turned out to be the wing center section which was envisaged as a mounting platform for all sorts of things, including cannon.

The design was assigned the designation XP-82, and it received a Factory Charge Number of NA-120. NAA was given the financial authorization to construct four prototype airframes under Contract Number AC-2029 in early February, 1944. Before the month ended, they received the go-ahead to build 500 P-82's.

Only two examples of the XP-82 were actually constructed, bearing Factory Construction Numbers 120-43742 and 120-43743, which were USAAF 44-83886 and 44-83887, respectively. These two examples were powered by two Rolls-Royce designed V-1650 engines, sub-designated as V-1650-23 and -25, to differentiate between the right and left hand engines, since they had counter-rotating propeller installations. The only real difference between these engines was in the crankshaft, cam shafts and electrical timing mechanisms, all of which were interchangeable.

To follow the XP-82's, two XP-82A's took the place of the second pair of authorized XP-82's. These latter two aircraft correspondingly bore Factory Construction Numbers 120-43744 and 120-43745, USAAF 44-83888 and 44-83889. These aircraft differed from the first two Twin Mustangs by the installation of General Motors, Allison Division's, V-1710-119 engines, without the benefit of counter-rotating propellers. Only the first example of the XP-82A was actually completed, however.

Aircraft construction was straightforward, utilizing established techniques. The basic design stemmed from the fuselage of the NA-105, XP-51F, which was the ultra-lightweight predecessor of the NA-126, P-51H. There were no major design revisions to the XP-51F from the radiator section forward. The after fuselage radiator shutters were extended 5.4 inches. The center wing construction was of constant chord and was 16 feet long, as was the horizontal stabilizer. The outer wing panels were, on the prototype XP-82's, the same as those found on the NA-126, as they were of heavier construction than those of the NA-105. The engine nacelles were identical to those used on the XP-51F, G and H. The canopies were shorter than those found on the lightweight F, G and J Mustangs, being identical to those on the P-51H. These factors gave the viewer the impression that the XP-82 was actually two P-51H fuselages joined together, rather than F's which gave rise to the erroneous opinion as to the aircraft's heritage. The XP-82 actually preceded the P-51H by four and a half months.

Even considering the accelerated development of war materials during WWII, the XP-82 design got off to a slow start. Apparently, there wasn't much "push" from the War Department. It is doubtful that the aircraft would have been available for combat in the event the island-hopping campaign had failed, nor would it have been available for the anticipated invasion of Japan. Part of the delay can be attributed to the increased emphasis placed on the modification of later models of the P-51, including a turboprop version. Only a few engineers were assigned to the development of the P-82. During actual construction, most of the delays arose through the build-up of the aircraft's wing and problems with the engines. The wing itself had to be modified from its original function of supporting a single engined fighter into something which would support a myriad of functions. The original wing panels were designed to mount three .50 caliber machine guns and carry one 1,000 pound bomb, or corresponding weight in the form of an external fuel tank, along with three to five 5-inch aircraft rockets. P-82 modifications included the removal of the gun bays and their associated ammunition trays, strengthening the wing spars and skin so that two 1,000 pound bombs could be carried, or in lieu of them, ten rockets or two fuel tanks with a capacity of up to 310 gallons each. The original wheel wells for the main landing gear had to be covered over and re-stressed for aerodynamic loads as well as for takeoff weight at the critical wing/fuselage junctions.

The XP-82 engines, although of proven design, required modification and testing. They were actually V-1650-11's which had been developed for the canceled P-51L program. They were derived from the Rolls-Royce R.M 14 S.M. engines used in the XP-51G. The -11's featured a Stromberg speed density carburetor which took care of the fuel/mixture problem for the pilot, and automatically increased the critical altitude of the engine. At 4,000 feet, with water injection, these engines were capable of producing a phenomenal 2270 BHP for a short period. The left engine required a slight modification to change the propeller's rotation, permitting the two propellers to rotate toward one another, which resulted in the elimination of the torque factor.

tion stage and then dropped. Two center section bomb racks were each capable of carrying a 1,000 pound bomb and were attached inboard of the wheel wells to supplement the racks on the outer wing panels. This would have permitted the P-82 to carry up

An F-82B with proposed air-to-ground armament. 500 pound bombs on the outer racks, 5" HVAR's on the inner ones, and the NAA developed with .50 caliber machine gun package attached under the wing center section, Unfortunately, the gun package was a "one off" experiment, as it would have been a handy thing to have had around during the Korean War.

The prototype P-82's were designed around several versatile concepts, not all of which were realized. As a pure interceptor version for air defense, there was one proposal that the right fuselage cockpit be completely faired over. The entire windscreen and canopy were to be removable and a single aluminum fairing was to be inserted in its place, which made for a clean fuselage and single pilot operation.

An attack fighter was another. This proposed version featured either an eight .50 caliber machine gun package installed in a center pod below the wing's center section, or a pod containing a 40 mm cannon hung in the same position. The cannon pod never got beyond the design stage, but the machine gun pod did and it carried 400 rounds per gun. Along with the six .50 caliber machine guns carried as normal armament in the wing's center section, the "attack" P-82 could lay down a withering fusillade. NAA's Chief Aerodynamicist, Ed Horkey, developed this package and its versatility was proven at the USAF's Eglin Proving Ground. Unfortunately, it never did reach production. It would have been quite valuable to the United Nations war effort in Korea had it become available.

A fighter-bomber version was another proposal. This too was carried through to the practical applica-

to 4,000 pounds of mixed bomb loads – and not exceed gross weight limits. The combat range of this version was 1,500 miles on the internal fuel alone, and there remained the provision or carrying external fuel on the outer bomb racks, if the bomb load was reduced by carrying lighter weight ordnance. In addition, the bomb racks were wired to carry trees of five 5-inch High Velocity Aircraft Rockets (HVAR's) in lieu of bombs on each rack, which permitted the "fighter-bomber" P-82 to fire an awesome barrage.

Horkey and his design team came up with two slightly different proposals for long-range escort missions. The first was a derivative of the fighter-bomber, which replaced the center section bombs with external fuel tanks. Together with the P-82's internal fuel capacity of 600 gallons, this version had a forecast range of 5,000 miles. A slight modification of this proposal moved these external tanks to the outer wing panels, and thus permitted the P-82 to carry four 310 gallon tanks. The range could be further increased by the installation of bladder tanks behind each cockpit. At an estimated speed of 240 TAS, the range was estimated to be 7,000 miles. This version later bore fruit as the P-82B, "Betty Jo," which set the speed record from Hawaii to New York City. As a long-range escort fighter, without

Setting out for the long haul. Lt. Col. Robert Thacker and Lt. John Ard depart Hawaii for New York. Total fuel for this mission was 2,215 gallons. The plan was to drop two of the 310 gallon tanks into the Pacific Ocean after they had been emptied, and the other two over an unpopulated area of Wyoming. Only one of the tanks could be jettisoned, - NAA

external appendages, the P-82's range on internal fuel approximated 2,500 miles or an average of nine and a half hours flying time. This is the version which the USAAF was really after for the Pacific Theater. They would not receive this version until long after the war was over at which time it was exemplified by the P-82E.

The last proposed version was the "night fighter," which later became known as the "all-weather interceptor." Two P-82B examples were used for development of these versions, becoming designated as the P-82C and D, the difference being in the type of radar equipment installed. The design featured a centerline pod which housed the heavy radar antennas, magnetrons and associated hardware, and it was pressurized by a nitrogen bottle for high altitude interceptions. This effort proved to be the most useful development of the Twin Mustang. When they became operational, with Allison engines, they were known as P-82F's, G's and H's.

One other version deserves mention, even though it was a one-off, in-the-field derivative of the P-82B. It was locally known, if not officially recognized, as the RF-82B. During the postwar years the need for a high altitude, long-range photo reconnaissance aircraft was recognized, but none of the existing aircraft seemed to meet the requirements. Development costs were prohibitive when it came to funding an aircraft solely for this function in peacetime military budgets. With the loss of the experimental Republic XF-12 and Hughes F-11, the two aircraft types which had originally been proposed for this function, there was a void to fill. A research group at Eglin Field decided to see if a P-82 might meet the need. The 3200th Proof Test Group, under the command of Colonel Merle Parks, reconstructed a pod which had been designed by NAA engineers, but was rejected due to instability. The new pod contained stations for up to seven cameras, including a K-22 with a 40 inch focal length. It proved to induce only a three mile per hour speed penalty, compared to a 15 MPH penalty with the radar pods on the P-82C/D. The RF-82B, 44-65172, was first flown on November 15, 1948 and was found to be stable and vibration free. However, after a year's worth of tests, it was relegated to use as a training mockup.

The first XP-82, 44-83886, was delivered and officially accepted by the USAAF on August 30, 1945. Prior to this date of course, NAA had been conducting their own tests at Los Angeles, California. The first flight had been made on June 16, 1945 with Engineering Test Pilot Joseph Barton (deceased) and Chief Engineering Test Pilot Edward M. Virgin at the controls – and it was a tale of what test pilots are made of!

For almost a week Joe Barton had been conducting ess turns, brake tests, high speed taxi tests and all the other necessary preflight ground work with the XP-82 before it became time for the actual first flight itself. Finally he was ready, and the next test was to determine at what speed the aircraft would lift off the runway. With the P-82 attaining this speed, the power would be pulled off, and preparations for the actual first flight would take place. Barton charged down the runway to the precomputed takeoff point and speed, and nothing happened! The P-82 remained solidly groundbound. He taxied back and tried it again, and this time he exceeded the estimated speed but it remained firmly planted on the runway. Once again, he tried, attaining even more speed and having to brake even harder to keep the XP-82 from running off the end of the 4,000 foot runway and into the Pacific Ocean. Still no liftoff, so he gave up for the day.

On the following day he asked NAA's famed P-51 Test Pilot, Bob Chilton, to ride along to help ascertain the nature of the problem. It was a repeat of the first day's frustrations; the Twin Mustang simply would not become airborne. Finally, on June 16th Barton got Ed Virgin to take the right seat and, after off-loading half the normal fuel load, they proceeded to the takeoff end of the runway. The plan was to get the aircraft just to the point where it was in the air, and then touch down and roll out to the end of the runway. Neither pilot wore a parachute. The acceleration began and, at the point where it should have begun to fly, it didn't once again so Barton pulled harder on the stick. Nothing! He

pulled even harder and still nothing, so he yanked it back and suddenly the Twin Mustang was both climbing and accelerating like mad. It was too late to get it back down on the runway safely. Virgin recalled that they flew around for an hour and a half getting the feel of the aircraft. It was considerably longer than normal for a first flight, but without parachutes they couldn't take a chance on a botched landing. The flight in itself was relatively uneventful, except for one potentially disastrous occurrence. A wing fillet tore free and wrapped itself around the horizontal stabilizer. As it turned out, the landing was successful and the aerodynamic team was given the task of determining why the XP-82 didn't want to take off. It took a lot of serious head scratching and a trip to the wind tunnel to find the cause.

The solution was simple. The XP-82 had been fitted with propellers which were reversed – with the left-hand propeller rotating counter-clockwise and the right propeller rotating clockwise. This resulted in the combined airstream being pushed from below to the top center of the middle wing section. In effect, it greatly increased the aerodynamic angle of attack of the wing's center section – to the point where it was producing drag instead of providing lift. The solution was to reverse the engines from their respective fuselages to the opposite side, and the problem was solved.

As we noted above, the USAAF accepted the first XP-82 on August 30, 1945, but returned it to NAA on a bailment contract that lasted for 18 months. Upon completion of airworthiness and stability flight tests by NAA pilots, the XP-82 was flown to Wright Field, Ohio for evaluation by Air Materiel Command personnel. It was re-designated as a YP-82 on December 13, 1947. On April 6, 1948, it was transferred to the National Advisory Committee for Aeronautics (NACA – now NASA) as NACA 114, to be stationed at Langley Field, Virginia. It was scrapped on October 5, 1955 after amassing 296 hours of flying time.

The second XP-82 made its initial flight on August 30, 1945 at a takeoff weight of 18,410 pounds, 1,400 pounds heavier than the initial flight of the first example. The flight lasted for an hour and five minutes but was marred by a considerable amount of roughness in the left engine. It, too, was accepted by the USAAF – on September 11, 1945 – and was returned to NAA for further flight testing. On March 18, 1946 it was returned to the USAAF as NAA's aircraft for use in the Official Performance Tests, which were to be conducted over the following two months. In October, 1947 it was transferred to NACA, this time to their Lewis facility at Cleveland, Ohio where it replaced a B-29 which was being used for high altitude air work. Eventually it was used to test Marquardt ram jet engines. It was destroyed in a ground accident on February 25, 1950 when the fuselages were twisted out of alignment as the aircraft slid off an icy runway. One of the fuselages of this aircraft still exists in a museum of sorts – in the Walter Soplata collection at Newbury, Ohio.

a power takeoff shaft that required a separate lubrication system. This was the same engine which was utilized on the XP-51J, and it had an infinitely variable speed supercharger, hydraulically actuated by an automatic dipstick control commanded by manifold as dictated by throttle position. The XP-51J was the one Mustang that Bob Chilton avoided entirely, apparently with good reason. This P-82A was turned over to the USAAF in October, 1945 and its disposal is unknown.

In spite of the engine shortcomings evidenced in the P-82A, the XP-82 had proven itself to be a viable design. Aircraft production now focused on the P-82B, which differed from the XP-82 solely by the

The production for the first order of F-82B's. The Twin Mustang with the number seven on the fin would be 44-65166. It went to SAC as a maintenance trainer, and was scrapped after attaining only 17 hours flying time! - NAA

Flight testing of the first P-82A, 44-83888, followed. Due to mechanical difficulties with the Allison V-1750-119 engines and the extremely critical pilot comments, the actual flight test program was dropped and construction of the second P-82A was discontinued. A major fault lay in the fact that the propellers of the "A" were not counter-rotating and the torque-induced imbalance was more than a pilot could physically handle. Another "negative" was that the V-1710 was a cantankerous engine at best, having an auxiliary supercharger impeller driven by

addition of a pressure carburetor on the Packard built Merlin V-1650-23/25 series engines. The original contract had called for 500 P-82A's but, with the postwar reduction of defense contracts, the number was reduced to the first 20 examples within the series. (For a period of time, the P-82B was designated P-82Z. The "Z" was used to designate groups of 20 aircraft which were contracted for to keep an aircraft production line open. As the P-51 line closed in November 1945, NAA would have had only their Navion and FJ Fury to work on, which meant that many of their talented people might be lost to other industries or to extended layoffs if NAA was forced to close down actual production of the P-82.)

Two of the P-82B's were modified into all-weather aircraft, although no one can remember exactly who came up with the idea of adding a radar

An F-82B loaded with Firebird air-to-air missiles which were manufactured by Ryan. They were tested at Holloman AFB, New Mexico. They were not intended for production. - USAF

The P-82B never reached the position of operational combat status, as the 20 examples were allocated to research and development or training roles. The first P-82B was flown on October 31, 1945 by Engineering Test Pilot George "Wheaties" Welch, one of the few USAAF pilots who managed to get their P-40's airborne over Pearl Harbor on December 7, 1941. Welch was subsequently killed on an F-100 test flight.

The P-82B's, serial numbered 44-65160-65179, remained on the Air Force roster through December, 1949 when they were withdrawn from service. In addition to being utilized by NAA for flight development, they were used by the Air Proving Ground at Eglin Field, Florida and by NACA. The P-82B had an odd fate for an aircraft contracted for in 1944. The first example was not officially accepted by the USAAF until January, 1947, even though production of this version had ceased in March, 1944. Actually, the first two were accepted in January, 1947, four more in February and 13 in March (figures including the P-82C and D). The last example of the P-82B was retained by NAA for static tests.

antenna and converting the co-pilot's cockpit into one for a radar operator. It may have been Ed Horkey, but even he couldn't remember. He was quick to point out that the P-82 was more of a group effort than previous NAA designs. Aircraft 44-65169 became the P-82C-1NA, while 44-65170 was the P-82D-1NA. The two aircraft differed only by the type of radar equipment installed in the aircraft and the shape of the radar pod, called the "dong." The P-82C carried an SCR-720 radar set that was an offshoot of the RAF Mk10 AI (Airborne Intercept) radar that had been carried by the Mosquito, Wellington, and the night-fighter Meteor NF11 and 13, as well as the P-61A and B. It was manufactured in the U.S. by the Western Electric Corporation. The P-82D mounted an APS-4 radar, which was one of U.S. design. Both the P-82C and D carried their radar antennas and associated equipment in the pressurized pod that extended from the wing's center section to a point just behind the propeller's arcs in the case of the C, or in front of the arcs in the D version. The nose of the pod was molded fiberglass, while the rest of the pod was of aluminum construction. As ungainly as they appeared, they actually created little additional drag. The pod was stressed to withstand 7.33 G's as was the entire aircraft, which was criteria at the time. Replacing the co-pilot's position was that of the radar observer (R/O). His instrument panel now carried only rudimentary flight instruments, airspeed, altimeter, artificial horizon. It mounted two cathode ray tubes; a B scope that gave range and azimuth and a C scope which gave azimuth and elevation. The pilot's cockpit also had a small radar scope which was slaved from the radar operator's and could provide him a choice of either the B or C presentation.

that of the C, did give a slightly higher airspeed. Additionally, since the D pod extended beyond the propeller arcs, there was less interference generated to the radar antennas. The D pod was selected as the design for the future all-weather P-82's.

In 1948 the P-82C was re-designated as the ZP-82C (the Z in this case denoting Obsolete. The P-82D was re-designated as an EP-82D with the E denoting "Exempt from Technical Orders"). Before the year ended, the ZP-82C was scrapped, along with the P-82B's that were assigned to Eglin Field. The EP-82D was transferred to Shepherd Field, Texas for use as an instructional airframe.

On NAA's flight line, F-82F's are prepared for test flights prior to receiving their gloss black paint schemes. Modelers should note that the propellers rotate toward the radar pod. - Esposito

The first flight of the P-82C occurred on March 21, 1946 and was piloted by Edgar A. Stewart, an NAA executive. Production Test Pilot Wilcox "Tex" Wild acted as his radar observer. The P-82D had its debut on August 30, 1946, being flown only by its pilot, Engineering Test Pilot George Krebs. In both instances the flight proved that, although there was some reduction in speed, there were no adverse aerodynamic characteristics generated by the appendages. The pod on the D, being less bulbous than

The next model of the Twin Mustang was the P-82E, which was the first version to see actual mass production and assignment to a combat unit. Essentially, the P-82E, re-designated the F-82E on June 11, 1948, was the same aircraft as the P-82B with the exception of being re-engined with Allison V-1710-143/145 engines. With the exception of the exhaust stacks and the carburetor intakes, they were identical in external appearance.

The use of Allison engines in the "E" came about

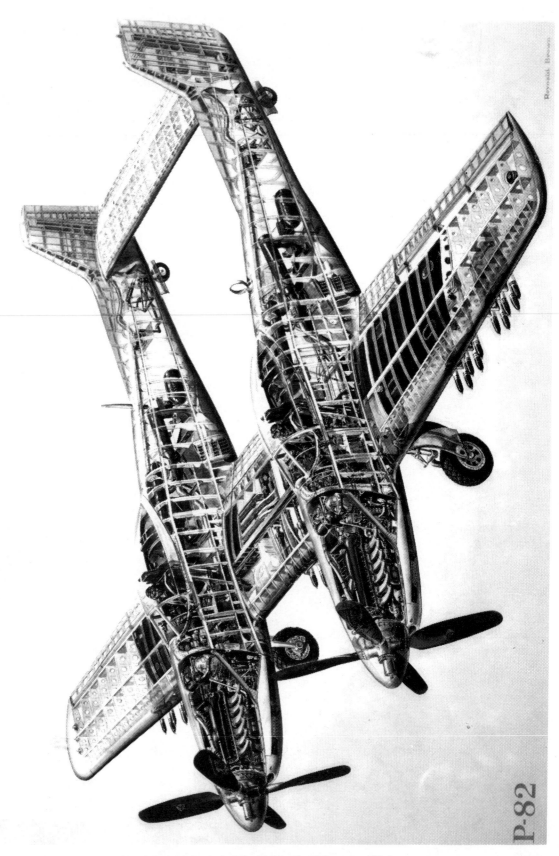

Cutaway drawing of an F-82B as drawn by North American Aviation artist Reynold Brown. - NAA

as a result of the Packard and Continental factories being closed due to post-War cutbacks. They produced the Rolls-Royce Merlin engines in the U.S. under license during WWII. If the P-82 was to continue with Merlin engines, these engines would have to be manufactured in England. For both economic and political reasons, the USAAF decided to allot the '82 engine contract to Allison.

Nevertheless, contract negotiations were conducted in August, 1945 for these engines, and although they met the test stand criteria of 150 operational hours between failures, they did not exceed this figure by much before they started to cause problems. The engines were difficult to keep in tune, due partially to the V-1710-145 being timed to run backwards. It didn't help things when you consider

Engine installation on the first F-82E, 46-255. An Allison V-1710-143 was installed in the left fuselage, a V-1710-145 in the right. They weighed 1595 pounds each, and produced 2250 bhp with water injection. At normal cruise power their output was 1100 bhp. - NAA

Some engine components, such as heads and blocks, were surplus Allison stock from earlier engines that Allison had left over from wartime production. The government had agreed to purchase the parts and then turn them back to Allison for the production of P-82 power plants. These engines were based on the same V-1710 engines which had been used in the P-40, P-38 and P-51A. They did not enjoy the fine reputation of the Merlin, being known by their pilots and ground crews as the "Allison Time Bomb" – for their propensity to quit at just the wrong time.

the inexperience of Air Force mechanics in the post-War period. Add to this the fact that the sparkplugs constantly fouled and, in the early engines, the plugs needed changing after every flight! Additional problems were backfiring while at both high and low power settings, along with rough running or surging. Although improvements resolved these problems to some degree, they were never entirely cured.

Oil leaks were another situation which almost defied a cure. The Allisons experienced: blow-by past the piston rings, venting oil overboard through the crankcase breather and having the oil froth in the

Cockpit modification to the co-pilot's cockpit of the XP-82 assigned to NACA - NASA via Layman

crankcase or in the oil header tanks. The problems appeared insurmountable – to the point where NAA was ready to throw in the towel on the entire P-82 program. A joint effort between the USAAF, NAA and Allison finally reduced the problems with the engines to a point where they were considered to be reliable enough for operational use.

A contract for 250 P-82E's was let on December 13, 1945, six months after the first flight of the P-82B, under the NAA designation NA-144. The actual procurement contract was not signed until October 10, 1946, however. This accounts for the fiscal year serial allocation of 46-255 for the first example. The contract itself was for 35 million dollars, which in the postwar years was a hefty windfall for the firm. Of this figure, $17 million was for the first 100 aircraft, $14.5 million for the second 150 aircraft, and $3.5 million for tools and associated equipment. The total cost for each P-82E and subsequent versions was $215,154. Cost overruns eventually exceeded the figure of $50 million, and to preclude further overruns, the contract for the P-82E's was reduced to 100 aircraft.

The first flight of the P-82E was made on April 17, 1947 by test pilot George Welch. The flight was troubled due to the engines throwing oil and backfiring. Additional static engine testing was forecast to delay the program extensively. The fear was borne out, as it took several months to get the Allison engines to a reliable standard. For some odd reason, the first four P-82E examples were unofficially redesignated as P-82A's by the USAAF during this period. They were at first restricted to engine tests, and then became P-82E's again when the testing program was terminated in June, 1948. This was one month after the remaining P-82E's were assigned to the 27th Fighter Escort Group.

Despite the engine problems, even as during the war – **NOTHING STOPS THE LINE!** As the F-82's rolled off the line at the NAA Inglewood plant they were towed to the Consolidated-Vultee Aircraft Corporation ramp at Downey, on the other side of the Los Angeles Airport. They were stored there until reliable engines were delivered from the Allison plant at Indianapolis, Indiana. The first 200 engines were accepted from Allison only to prevent slippage of the P-82 program, but they could only be run at reduced power settings. A production line was set up

on the ramp at Downey, which was cheaper than moving the aircraft back across the airport and kept the aircraft from getting in the way of the emerging FJ/F-86 program. All of this was not inexpensive and NAA had to cover the expenses, security, insurance, etc. These additional fees ran into millions of dollars and slipped the whole program approximately two years. They practically negated the profit value of the aircraft. By the end of 1947, 130 P-82's were in storage on the Downey ramp.

The first P-82E (A) was accepted by what was now the United States Air Force – in September, 1947. (The USAF became a separate branch of the military on July 26, 1947. At this time, all USAAF Fields were re-designated as Air Force Bases.)

The first and second P-82E's were retained by NAA for further flight evaluation, while the remainder were prepared for assignment to other units. The first of the P-82E's designated to become operational with a combat unit was accepted in January, 1948 and by June, the USAF had received a total of 72, which closed out the fiscal year (and satisfied the comptroller's paperwork.) Twenty-two were accepted in July, 1948, now designated F-82E, effective June 1, 1948. The last two were delivered in October and December. All of the combat operational aircraft (Very Long Range) went to the Strategic Air Command's 27th Fighter Group (VLR) at Kearney AFB, Nebraska, which had a total of 81 aircraft on their roster by the end of the calendar year. *(See Chapter Two)*

There were three production versions of the all-weather F-82 with the only major difference between them being the type of radar equipment installed. In the case of the F-82H, additional modifications of the F-82F systems were made – for winterization. Flight tests of the F-82C/D had proven that the F-82 would make an adequate all-weather fighter, with some reservations. It was stiff on the controls, hard to slow down behind an intercepted target as it had no speed brakes (why this addition was not fitted remains a mystery, for NAA did have experience with these when they built the A-36 Invader (P-51A Mustang) and the addition should not have been a that great a problem.) It also had poor visibility for night flying. A good deal of the night visibility problem was alleviated by modifying the engine exhaust stacks with flame dampening ports.

To reduce the heavy handling, hydraulic boost systems were installed, with pumps for the ailerons and additional pumps in each fuselage for the elevator and rudders, the latter being connected by cables in the event that pump failure or battle damage would render one system inoperative.

There was little doubt that the all-weather Twin Mustang was to be nothing more than an interim fighter as far as the USAF was concerned. Proponents had a hard time selling the concept to the AF, as the Air Materiel Command had recommended that the all-weather versions should be converted back to escort fighters, due to their inherent problems.(!) A design competition was taking place with Northrop, Curtiss and others vying for contracts for all-weather jet fighters. The F-82 would have to fill the gap between the newer jets and the aging P-61 or there would not be any all-weather protection available for several years.

Production of the F-82F, NA-149, commenced with the termination of the F-82E and the line ran simultaneously with the F-82G, NA-150. The first of the all-weather F-82's to be flown was actually F-82G 46-355 which was test hopped by NAA Engineering Test Pilot Nicholas Pickard on December 8, 1947. The first F-82F was not flown until March 11, 1948.

All F-82 production was completed by April 30, 1948 and the NAA production line gave way to the F-86. At this time all of the additional F-82 work was being accomplished on the Convair ramp at Downey – and it took some effort. In the lengthy period between construction and acceptance, the aircraft sustained corrosion and other related problems while the Twin Mustangs were in open storage. Finally, with the supply of viable engines catching up with the airframes, deliveries could take place. The first F-82G was accepted by the USAF in February, 1948, while the remaining all-weather examples were delivered in fiscal 1949.

The first 29 F-82F's were delivered to the five squadrons that comprised the 52nd and 325th Fighter (All-Weather) Groups and the 319th Fighter (All-Weather) Squadron, commencing in September, 1948 under Projects ADC 8-PF-21. CAC 8-PF-32 and CAC-32 respectively. These deliveries were followed by the F-82G's which were flown to the 347th Fighter (All-Weather) Group under project FAF-33, and then the remainder of the F-82F's to the 52nd and 325th F(AW)G's. The F-82H's went to Ladd AFB, Alaska for service with the 449th Fighter (All-Weather) Squadron under Project AAC 97655. By March, 1949 all of the all-weather Twin Mustangs had been dispatched to their intended destinations, probably to the considerable relief of the concerned NAA personnel!

* It should be noted the alphabetical reversal between the F-82F and G and the aircraft's serial numbers. The F-82G was serial blocked 46-355/383 and 46-389/404 (c/n 150-38241/38269 and 150-38275/38290) while the F-82F blocked 46-405/495 (c/n 149-38291/38381). The F-82F filled in the gaps in the serial number allocations, since they were modified F-82F's and G's, blocking 46-384/388 and 46-496/504 (c/n 150-38270/38274 and 150-38382/38390). The F-82H's were five modified F-82G's and nine F-92F's. The F-82H was first flown on November 15, 1949.

Chapter Two

On Long Range Escort

The F-82E was assigned to the 27th Fighter Escort Group, Strategic Air Command, stationed at Kearney AFB, Nebraska. The 27th FEG was a unit which had an interesting and varied history. It was originally constituted as the 27th Bombardment Group in WWII. The Group had originally been slated to fly the Douglas A-24 ''Dauntless'' (the USAAF's version of the Navy's SBD). The Commanding Officer, Colonel John N. Davies, along with 20 of his pilots were in Australia to accept these aircraft when the war started. The aircraft did not arrive on schedule, so Davies and his men were transferred into other outfits. The aircraft, when they finally arrived, were also dispersed among other squadrons. For the remainder of the men of the 27th BG, who were awaiting their aircraft at their home station on Luzon, the war took a critical twist and most of them wound up fighting as infantry on Bataan and Corregidor. Unhappily, most of them were lost.

The 27th BG designation was returned to the U.S. and an entirely new group was formed around it. Revitalized, the 27th BG went to North Africa, Sicily, Italy and Southern France. They flew a succession of A-36's, P-40's and P-47's. The 27th BG was deactivated on November 7, 1945.

The 27th was soon reactivated as the 27th Fighter Group in Germany and equipped with P-47's. They remained there only a year, and were returned to the U.S. ''less personnel and equipment'' on June 25, 1947. The 27th was relocated to Kearney Field on July 16, 1947 where they were assigned to the emerging Strategic Air Command. They were redesignated as the 27th Fighter Escort Group.

At this time the 27th FEG was led by Colonel Edwin A. Doss who was succeeded five month later by Colonel Ashley Packard. Packard was replaced by Colonel Cy Williams in March, 1948 and he was the primary group commander during the group's F-82E era.

The 27th FEG began with 55 officers and enlisted men who had been transferred to Kearney from the 82nd Fighter Group (Very Long Range) at Grenier Field, New Hampshire. Their first mission aircraft was the P-51D Mustang. Expansion brought the reactivation of three fighter squadrons; the 522nd, led by Major Virgil Meroney, the ex-commanding officer of the 352nd FG (10 victories); the 524th, led by Lt. Colonel George V. Williams and the 523rd, led by the legendary Colonel Donald Blakeslee (15.5 victories).

The military posture at the time the F-82's were delivered to the 27th FEG was one of full scale alert. Stalin had starting rattling sabers over Berlin in late winter 1947 – and their clatter kept getting louder. All of the USAF's fighter units were placed on ever increasing alert status. At this time SAC had only one operational VLR group, and that was the 82nd at Grenier. They had F-51H's, and they were soon dispatched to Alaska to assist in the defense of that area. This left SAC with no aircraft capable of escorting their own bombers to anywhere they would actually require an escort. (The P-51D was not considered to be a VLR aircraft, in spite of its proven long legs, and the few SAC F-80's and F-84's were obviously short legged in that era.)

The 27th FEG was transferred to McChord Field, Washington with F-51D's in late January 1948 where they stood strip alert during daylight hours. The 325th Fighter Group's F-61 Black Widows took over the function at night. There was a lot of talk among the crews as to who was going to do what if ''the balloon went up.'' It wasn't all idle talk, since over 70% of the men composing the 27th FEG had seen action during WWII. When the F-82's started to arrive the Group's situation changed, for they had to cross-train into twin-engined aircraft whose configuration itself was suspect. However, NAA had spent so much time in the development of the aircraft that all of its technical aspects and characteristics were well known. In addition, the Group was supported by NAA Technical Representatives (Tech Reps). The transition training was carried out without undue delay.

attacks across the continental United States. Mission lengths could run between eight and ten hours. Since the F-82's had both an autopilot and a co-pilot they were not as fatiguing for their pilots as they would have been in a single seat fighter. The largest concern for the Twin Mustang crews was the tumultu-

F-82C
Pilot's Cockpit - Forward View

1. Clock	2. Airspeed indicator
3. Remote indicating compass	4. Turn and bank indicator
5. Generator ammeters	6. Directional gyro
7. Rate of climb indicator	8. K-18 gun sight
9. Manifold pressure drain	10. Voltmeter
11. Artificial horizon	12. Engine air temperature
13. Left engine gage	14. Suction gage
15. Right engine gage	16. Hydraulic boost gage
17. Tachometer	18. Coolant temperature gage
19. Manifold pressure gage	20. Heat control panel
21. Emergency L.G. release	22. Parking brake handle
23. Defrost Control	24. Engine Control Panel
25. Bomb/rocket release	26. Emergency canopy release
27 Surface control lock	28. Fuel control panel
29. Cockpit light rheostat	30. Hydraulic press. ind. light
31. Landing gear indicator	32. Warning horn cutout sw.
33. Armament control panel	34. Accelerometer
35. Altimeter	36. Radio compass indicator
37. Water injection switch	

Mission profiles were flown with the F-82E's being called upon to escort B-29's and B-36's on mock

F-82F
Pilot's Cockpit - Left Side

1. Anti G suit valve	2. Fuel valve circuit breaker
3. Voltmeter selector	4. Voltage regulators
5. Armrest/std. checklist	6. Wing flap handle
7. Taxi light switches	8. Landing light switch
9. Aileron trim tab control	10. Remote compass corr. card
11. Propeller controls	12. Throttles (Sight range ctrl.)
13. Radio push-to-talk sw.	14. Oil radiator switches
15. Flap emergency switch	16. Coolant radiator switches
17. Bomb sequence switch	18. Bomb salvo switch
19. Throttle & Prop frictions	20. Gun nacelle emergency rel.
21. Gunsight selector/dimmer	22. Landing gear control
23. Elevator trim tab control	24. Rudder trim tab control
25. Landing gear instructions	26. Circuit breaker panel
27. Anti G suit connection	28. Heat and vent outlet

ous weather over the Great Plains during the winter blizzard or summer thunderstorm seasons. It became common practice to launch the entire group in virtually zero-zero visibility rain or snow conditions in order to meet a mission commitment. This was something unheard of in single engined fighters. But with one pilot able to go on instruments while sitting on the end of the runway while the other pilot lined up the aircraft and started the takeoff roll visually, the fighter could be taken off quite safely. These types of missions would not be attempted in today's climate of safety consciousness, but in 1948-49 they were considered a necessity for training.

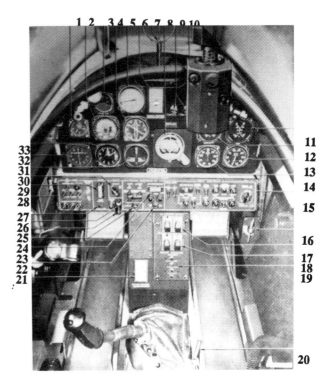

F-82E
Pilot's Cockpit - Right Side

1. Hydraulic boost switch
2. Keying switch
3. Fuselage light switch
4. Position light switches
5. Command radio trasf. sw.
6. Canopy hand crank
t. SCR-695B radio panel
8. AN/ARC-3 radio panel
9. AN/ARC-6 radio compass
10. Oxygen flow indicator
11. Oxygen pressure indicator
12. Rocket firing control
13. Oxygen regulator
14. Microphone cord
15. Circuit breaker panel
16. Heat and vent outlet
17. Data case
18. Cockpit air volume control
19. Aileron boost circ. breaker
20. Drop tank selector
21. Free air temperature gage
22. Emerg, coolant air flap rel.
23. Rudder pedal disconnect
24. radio circuit breaker panel

F-82E
Co-pilot's Cockpit - Forward View

1. Clock
2. Airspeed indicator
3. Free air temperature gage
4. Remote indicating compass
5. Accelerometer
6. Remote compass corr. card
7. Ring and bead sight
8. Artificial horizon
9. Manifold pressure drain
10. Gun camera
11. Oil temperature gage
12. Coolant temperature gage
13. Tachometer
14. Manifold pressure gage
15. Heat control panel
16. Emergency landing gear Rel.
17. Oil radiator switches
18. Coolant radiator switches
19. Circuit breaker panel
20. Emergency canopy release
21. Standby compass card
22. Fire indicator test switch
23. Fire extinguisher switch
24. Landing gear position light
25. Bomb salvo switch
26. Mixture control switches
27. Gunnery control shift sw.
28. Fuel control panel
29. Gun selector switch
30. Cockpit light rheostat
31. Turn and bank indicator
32. Radio compass indicator
33. Altimeter

Since the 27th FEG was the only VLR group in the USAF, they never knew when they might be called upon to fly for "the real thing" and the F-82E was the only fighter in the inventory that was capable of escorting bombers from England to Moscow (although it was doubtful that only one fighter group would have made much of an impact if it had come to that!)

The 27th FEG started flying long range overwater missions in late January 1949 as a part of their training. The group would take off from Kearney and fly to McDill AFB, Florida to refuel. They then proceeded to Ramey AFB, Puerto Rico to refuel again. Next stop was Howard AFB, Panama, and then it was on to Jamaica for more gas. The final leg took them back to Kearney. These missions would encompass a couple of days, but it was sound navigational training and endurance testing for both pilot and aircraft.

Kearney was closed (as a major Air Force Base) in March, 1949 and the 27th FEG relocated to Bergstrom AFB, Texas on March 11, 1949. The mission remained the same and the VLR flights continued to Puerto Rico. One notable mission was accomplished nonstop (!) between Bergstrom and Ramey in ten hours, covering a distance of 2,300 miles. This was a feat unequaled by any other fighter type until aerial refueling became commonplace.

The 27th FEG started transition training into the F-84E in the spring of 1950. By August 3 the group

F-82E
Co-pilot;s Cockpit - Left Side

1. Anti G suit valve
2. Heat and vent outlet
3. Anti G suit connection
4. Armrest and std. checklist
5. Landing light circuit breaker
6. Landing light switch
7. Taxi light switch
8. Throttles
9. Radio push-to-talk switch
10. Propeller controls
11. Wing flap emergency sw.
12. Nacelle emergency rlease
13. Rudder trim tab control
14. Elevator trim tab control
15. Rudder pedal disconnect

F-82E
Co-pilot's Cockpit - Right Side

1. AN/ARN-6 Radio compass Control panel
2. Command radio transfer switch
3. SCR 559B radio control panel
4. AN/ARC-3 radio control panel
5. Canopy handcrank
6. Oxygen flow indicator
7. Oxygen pressure indicator
8. Oxygen regulator
9. Microphone cord
10. Heat and vent outlet
11. Data case
12. Cockpit air volume control
13. Emergency coolant air flap release
14. Rudder pedal disconnect

was considered as combat ready in the Thunderjet. The F-82E's were then flown to Warner-Robbins AFB, Georgia to await a decision as to whether they

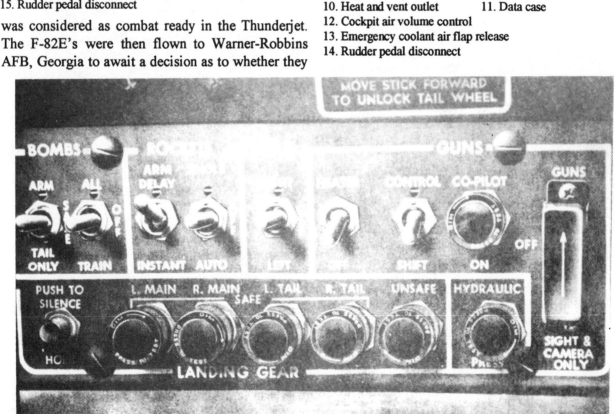

Armament control panel - F-82E

Pilot's engine control panel F-82E

would be scrapped, transferred to the Air National Guard, the Air Force Reserve or maybe to the Far East for service in Korea. Despite Lt. General George E. Stratemeyer's request for 21 additional F-82's for his Far East Air Force (made on June 30, 1950), it was decided to scrap all the F-82E's. Apparently the AF decision was taken for logistical rea-

sons. The disposal of the Twin Mustangs took place through mid-1951, with engines and other usable parts being shipped to the Korean war zone or to Alaska. The 27th FEG, now under the command of Don Blakeslee, fought extensively in the Korean skies – in their new jets.

Chapter Three

Surprisingly, the most extensive use of the F-82 was in the air defense role. Painted a shiny black, radar equipped '82's stood alert – when they could be kept in commission!

Twin Mustangs for Defense

The 325th Fighter Group

The first F-82F's were delivered to the 325th Fighter (All-Weather) Group in 1948. The 325th had been a fighter group which had fought with the 12th and 14th Air Forces during WWII, both in North Africa and Italy where they flew P-40's P-47's and P-51's. They were deactivated upon their return to the U.S. on October 20, 1945. The 325th was reacti-

vated on May 21, 1947 and was re-designated as the 325th Fighter Group (All-Weather) at that time.

Their reactivation took place at Mitchell Field, Long Island, New York but it was only a "paper assignment." The designation was transferred to Hamilton Field, California on December 2, 1947 "less personnel and equipment." During WWII, the 325th

"Upstairs Maid" from the 319th F(AW)S, Moses Lake, Washington. - Paul Kajor via Menard

"Spokane Spook" another F-82F from the 319th F(AW)S - Paul Kajor via Menard

had been composed of the 317th, 318th and 319th Fighter Squadrons and eventually, the group would regain these same three squadrons.

Initially, however, only the 318th Fighter Squadron was attached, and again, it was only a paper assignment, awaiting personnel and equipment. The first actual squadron to join the 325th FG (AW) was the 425th Night Fighter Squadron which had been on paper at March Field, California. (The 425th was the first squadron assigned to the Air Defense Command when it was formed in July 1946. ADC attached the squadron to the 4th Air Force. The squadron had one P-61 at the time and few men).

The 4th Air Force transferred the 425th NFS to McChord Field, Washington on September 1, 1946.

"Night Flight" from the 319th F(AW)S, it's a shame that the artist is unknown - Paul Kajor via Menard

On August 25, 1947 the 425th NFS was re-designated as the 317th Fighter Squadron. When the 325th FG (AW) established themselves at Hamilton Field they were cadried by men of the 317th FS who had been transferred from McChord on November 24.

The squadron relocated to France Field, Canal Zone on January 4th, 1948 and flew the Black Widows from there until late December when the F-82's started to arrive under Project Caribbean Air Command 8-PF-32, dated October 19, 1948. (The P-61's, by the way, were scrapped, but rumors persist that

An F-82F of the 318th F(AW)S snuggles up to its wingman near Moses Lake, Washington - Dickman via Davis

As soon as an adequate number of men were on board within the 325th FG (AW), the 318th Fighter Squadron (All-Weather) was activated, on August 21, 1947. The official activation of the 317th FS (AW) followed four days later.

In the spring of 1948 both squadrons transitioned from P-61's onto F-82F's (the actual re-designation from P to F was mandated June 11, 1948). When the two squadrons were considered combat ready they relocated once again. The 317th FS (AW) moved to Moses Lake AFB, Washington (later redesignated Larson AFB) on November 26, 1948. The 318th FS (AW) moved back to McChord to absorb the detachment which had been on air defense duties with P-61's.

The 319th Fighter Squadron (All-Weather) in the meantime, had been reactivated as a result of Air Force policy that did away with many of the higher numerical squadron designations. On September 1, 1947 the 414th NFS was redesignated the 319th FS (AW). The 319th FS (AW) was attached to the 6th Fighter Wing, 6th Air Force, Caribbean Air Command at Rio Hato, Panama where they flew P-61's.

some of these aircraft are scattered among various South American countries.)

With an AF determination that all-weather fighters were no longer needed in Central America, the 319th FS (AW) was transferred to McChord AFB, effective May 12, 1949. The actual movement of aircraft began in March with all but the last two Twin Mustangs having flown north by the end of April. The last two arrived at McChord on May 12. This last movement for the squadron was under the authorization of Continental Air Command CNC-107. (The CNC, better known as CONAC, was the parent unit of the Air Defense Command, which actually did not become an autonomous command until January 1, 1951.)

The 319th FS (AW) was assigned to the 325th Fighter Group upon their arrival at McChord, but they only remained at McChord until September 2, 1949 when they relocated to Moses Lake AFB to share air defense alert with the 317th FS (AW).

The three squadrons of the 325th FG (AW) flew in the air defense role and they were spread mighty thin, considering their responsibilities. With only one all-weather fighter group located on the west coast of the U.S., there was not much protection to be offered in the event of an attack. The Royal Canadian Air Force and USAF day fighters were expected to take care of any threat that might occur from over the North Pole, but there was not all that much strength involved. In actuality, the threat was not as great as many politicians made it out to be, for the USSR had only the TU-4 (a plagiarized version of the Boeing B-29) for a strategic weapon – and not very many of them at that! Still, whatever enemy threat there was had to be considered. The industrial Northeast, along with the sensitive aviation industry in the Northwest, did feel vulnerable to enemy attack and demanded a show of aerial protection.

ing. They were replaced at Moses Lake by the 91st and 92nd Fighter Squadrons of the 81st Fighter Wing which relocated their F-86's from New Mexico. By January 1, 1951 the 325th F(AW)G was operating a mixed bag of 19 F-82's and F-94's. On May 1, 1951 the 325th was re-designated once again, as the 325th Fighter Interceptor Group, which was more in line with their new role, since it had been determined that the F-94 was not actually the all-weather fighter it had been made out to be. (As an interceptor it greatly exceeded the abilities of the F-82, through its improved fire control system, but it was not considered as an all-weather aircraft due to its inadequate de-icing system.)

From the middle of May, 1951 through the mid-

An indication of how tightly the canopies slide on their tracks. A 52nd F(AW)S F-82F, The F-82F weighed 320 pounds more than the F-82G, due to the type of radar equipment carried. The winterized F-82H weighed 300 pounds more then the F-82G, due to the double clamping of cooling and hydraulic lines, along with additional radio equipment and a heavier duty electrical system - USAF via Greenhalgh

As soon as the F-94A Starfire became available the 325th Fighter Group was slated for transition into these fighters, to appease the political pressure groups in the Northwest. On April 23, 1950, the 317th and the 319th moved back to McChord to join the 318th F(AW)S to commence conversion train-

dle of August the 325th FIG's F-82's were either transferred to the 52nd FIG, or to Brooks AFB, Texas where they were salvaged.

Over a period of years the 325th FIG was broken up, with the 319th FIG being sent to Suwon, Korea with F-94B's in February 1952. (They destroyed four enemy aircraft in night air-to-air combat.) The 318th FIS was dispatched to Thule, Greenland in June 1953, and the 317th FIS went to Alaska in 1957.

and equipment."

The 52nd FG reactivated at Mitchell Field on June 25, 1947 under the command of Colonel Oliver S. Cellini (who would later take over the 51st

Ground crews for the 52nd F(AW)G pose in front of Col. Cellini's F-82 at Nellis AFB, March, 1950. Note that the tail warning radar antennas are mounted on both sides of each vertical fin (installed in the big stars on the fin). - USAF via Greenhalgh

The 52nd Fighter Group

The 52nd Fighter Group (All-Weather) was based at Mitchell Field, Long Island, New York when they first received their F-82F's. The 52nd Fighter Group had flown the British Spitfire as their first operational aircraft during WWII with the 8th Air Force in England. They were then transferred to the 12th Air Force in Africa in November, 1942. They were, then, reassigned to the 15th Air Force in May 1944. They converted into P-51's and moved into Italy. The unit was de-activated in the U.S in November, 1945, only to be reactivated in Europe in 1946 under the command of Colonel Carrol W. McColpin, the ex-commanding officer of the 4th Fighter Group's 336th Fighter Squadron (12 kills). They were again returned to the U.S. in June 1947 "less personnel

Fighter Interceptor Group in Korea). Two squadrons of the original 52nd FG were also reactivated at this time, the 2nd and the 5th Fighter Squadrons. The third squadron which had originally been attached to the group, the 4th, which had been deactivated in 1945, was reactivated with the 347th Fighter Group (All-Weather) of which more later. *(See Chapter 5)*

At the time of reactivation, the 52nd FG (AW) was assigned P-61's for the defense of the U.S.'s eastern seaboard during darkness and inclement weather, sharing the air defense role with the 14th Fighter Group, which covered daylight hours in F-84's. The first F-82F's started to arrive in June 1948 and transition training from the P-61 began, but it was erratic at best. Where the P-61 had been a reliable, if slow and steady flying machine, the F-82 proved to the men of the 52nd FG (AW) that it was an absolute "maintenance hog." This was particularly true in respect to the radar systems, and the Allison engines. Exactly how many of these problems can be laid directly on the aircraft, and how many more on the inexperience of the ground crews and technicians cannot be determined. Official records indicate great dissatisfaction with the Twin Mustang during its first two years of operational life with the 52nd FG (AW). The 52nd was more outspoken than any other Group to operate the F-82!

The 52nd FG (AW) relocated to McGuire AFB, Ft. Dix, New Jersey on November 10, 1949 and continued to fly the F-82 from there until the 2nd FS (AW) started conversion training into the F-94A. The 5th FS (AW) had a similar change. The F-94 was introduced into the CONAC in August, 1950 and, although it was considered sub-standard in respect to the F-89 which was just starting to come off the production line, production of the F-94 was at a faster rate. The men of 52nd Fighter now (All-

Col. Cellini addresses a group of West Point Cadets visiting the 52nd F(AW)S in June 1949. - USAF via Greenhalgh

As of October 1, 1948, 22 F-82's were on the Group's roster, but their in-commission rate was a sparse 25%. Many of their problems were never resolved. For example: In January, 1950 the two squadrons totaled 7,500 man hours on aircraft maintenance, while the base shops used an additional 5,000 hours on the F-82F. On the average, it took 24 man-hours of maintenance effort for each hour of flight! They conceded that the F-82 could never perform 100% effectively under the existing personnel and supply conditions, both plagued by short-

Weather) Group Composite were elated with their new jets due to their incredible reliability and the ability to fly both faster and higher. The best altitude the 52nd F(AW)G could ever attain from their F-82's was 22,000 feet - according to their official history!

The 84th Fighter Wing

Paralleling the history of the 52nd Fighter Group is that of the 84th Fighter Wing (All-Weather). The 84th FW (AW) had one operational flying squadron, the 496th Fighter Squadron (All-Weather), and two others "on paper" – the 497th and 498th, which did not become operational until the mid-1950's when they were activated as first-line USAF squadrons. The 84th FW (AW) was a USAF Reserve Corollary Wing, which was a forerunner of today's Associated Air Force Reserve (AFRES) units of part-time citizen-soldiers. The 84th FW (AW) was stationed at Mitchell Field from the time of their organization on June 1, 1949 until they followed the 52nd FG (AW) to McGuire AFB. Under the corollary program, the 84th Wing did not actually possess any aircraft of their own, but the members of the wing took their weekend drill and summer encampments alongside the men of the 52nd FG (AW). Their Commanding Officer was Brig. Gen. Arthur L. McCullough.

The single flying squadron was faced with the same aircraft problems as the 52nd FG (AW). There were just too few available operational aircraft for them to train with. The few in-commission F-82's were usually required for the 52nd FG (AW)'s air defense mission. As of April 1950, only one AFRES pilot had become certified in the F-82, and by June, only four more had managed to become type-rated. When the 496th FS (AW) held their summer encampment those five pilots did manage to get in some 226 hours flying time in the Twin Mustang, but this figure was well below the time allotted to them by the 52nd FG (AW). Also unfortunate was the fact that the squadron was never able to qualify any radar operators. Those that were with the squadron when they had been associated with the P-61 elected to drop out of the AFRES when the F-82 came along.

When the Korean War started, the USAF decided that it could no longer support the corollary program, because of the activation of most of the Reserve squadrons and all of the additional effort involved in preparing for the war. The corollary program was curtailed in September, 1950. The 496th FS (AW) was de-activated at this same time, only to be reactivated at Hamilton AFB, California in 1952 as a part of the regular Air Force's Air Defense Command.

Chapter Four

Although Arctic temperatures didn't do anything to enhance the '82's in-commission rate, the Twin Mustang gained a new mission – as a tactical reconnaisance aircraft!

Over the Arctic – the F-82 in Alaska

The winterized version of the F-82, the F-82H, was delivered to the Alaskan Air Command's 449th Fighter Squadron (All-Weather) at Ladd AFB, Fairbanks, beginning on March 20, 1949. The first shipment was a flight of six aircraft, flown by squadron pilots. They picked them up from March AFB, California – from the modification depot where final winterization of the airframes had taken place. After acceptance checks, the ferry mission became airborne, only to discover that one of the F-82's had problems. The left outboard fuel tank wouldn't feed – and wouldn't provide sufficient fuel reserves for the planned non-stop flight to Ladd. With an escort of another F-82, the problem aircraft's pilot elected to head for Alameda Naval Air Station where, on arrival, he discovered that the left landing gear would not extend. The gear was lowered using the emergency system and the aircraft landed without further difficulty. These two aircraft did not arrive at Ladd until April 29.

The 449th FS (AW)'s first reported F-82H accident occurred on April 11, 1949 when one aircraft was landed off to the side of the runway at Galena AFS, Alaska causing major damage. Meanwhile, the remaining aircraft allotted to the squadron were still being ferried north and, with the exception of one F-82H having to divert to Great Falls, Montana due to propeller problems, they all arrived without further incident. By the end of June, 1949 the squadron had all 14 of the F-82H version of the Twin Mustang on board, along with two B-25's and one C-47 for transportation and cargo missions.

The 449th FS (AW) had originally been designated as the 449th Fighter Squadron and had flown with the 51st Fighter Group in China during WWII. The squadron was deactivated on Christmas Day 1945. It was reactivated on September 1, 1947 when it was attached to the 5001st Composite Wing, commanded by Brig. Gen. Dale V. Gaffney. Their first mission aircraft were P-61's and, on receipt of the aircraft, they moved to Davis Field, Adak Island, midway down the Aleutian chain. The first Squadron Commander was Lt. Col. Leon G. Lewis. As of January, 1949 the squadron had eight P-61's and one B-26C for operational and hack flying.

The 449th FS (AW) ceased flying operations at Davis Field on February 29, 1949 to prepare for their impending move to Ladd Field. On March, 5 they were ready to depart but adverse flying conditions delayed the actual move until March 10th. Even then, they were caught in more bad weather and had to land at Thornborough Air Base. They finally arrived at the 26 Mile Base (later renamed Eielson AFB) 26 miles from Fairbanks on March, 11. The squadron's ground echelon was finally able to get out of Davis Field on March 23 and it took them five days to get to Ladd. Four of the squadron's P-61's were flown back to the U.S. where they were assigned to the 52nd FG (AW) to increase their complement of aircraft until they received more F-82's of their own, while the remaining 449th FS (AW) P-61's were scrapped in Alaska.

To assist in the squadron's transition training into the F-82, four p73 F-82E's and their crews were sent to Ladd from the 27th Fighter Escort Group. The 27th's crews weren't too enthusiastic about the impending TDY (Temporary Duty). While it was springtime in Nebraska, it was still very much winter in Fairbanks. After a month of hands-on training, the aircraft and crews from the 27th FEG were permitted to fly home, much to their delight.

For the type of all-weather flying necessary in

Alaska and the Aleutians, the F-82 proved to be the most suitable aircraft available in the USAF inventory. The aircraft's long-range capabilities permitted the fighters to range over the vast expanses of land and water without undue difficulty. If they encountered bad weather at their destination, they still had enough fuel reserve to seek out a suitable alternate. Their role in defending Alaska was quite an obligation. With the exception of the F-80's of the 57th Fighter Group, the F-82 represented the only fighter power in the territory and they received constant challenges.

although some GCI operators swore they were jets. In either case, the F-82's could never get close enough for positive identification. But, turn-about is fair play, and between May 5 and May 9, 1949, the 449th FS (AW) demonstrated to the Russians that their new F-82 was a different breed than the older P-61's. Four combat loaded F-82's launched on reconnaissance and surveillance missions over Russia from Marks Field. Altogether nine sorties were flown, but the effectiveness was deemed poor, due to the inexperience of the aircrews and the lack of suitable communications between the fighters and

An anomoly, is this an F-82E modified into an all-weather fighter, or an F-82H refitted wuth an E's cowlings and exhaust stacks? Although two F-83E's went to Ladd Field for cold weather tests in March 1949, they were only there for a week before returning to Eglin Field. There is no record of any F-82E's being assigned to the 449th F(AW)S. - Monaghan via Isham

Even though the 449th FS (AW) was headquartered at Ladd Field, they maintained detachments at Galena, Davis and Marks Fields, Nome. For operations at Davis, their Ground Controlled Intercept (GCI) site was located on Sheyma, at the far end of the Aleutian chain. When Russian aircraft were spotted on the Sheyma radar, which usually occurred once a week, an F-82 had to be scrambled on it. Apparently the Russians were monitoring the appropriate radio frequencies, for as the F-82 would near their position, they would turn and head for home. With their height and speed advantage they would quickly pull away. They were believed to be TU-4's

their GCI at Marks. Even so, they let the Russians know that it wasn't to be all their way.

Additional missions of another sort were also required. One of the major problems in Alaska during the spring months is massive flooding brought on by ice jams along many rivers bordering populated areas. The 449th FS (AW) was called on to assist the 57th FG in bombing these blockages. Between May 19 and May 28, 1949, a total of 23 ice bombing missions were flown, amounting to 51 sorties between the F-82's and B-25's. They dropped a total of 27 tons of bombs on the ice packs, which alleviated a lot of the flooding problems facing many downstream villages. These types of missions were excellent tactical training practice for the aircrews and ground personnel and they would be scheduled each year during the late spring and early summer months.

The first Operational Readiness Inspection (ORI) for an F-82 unit took place between June 6 and June 9, 1949, and the 449th didn't fare too well. Only 50% of the squadron's aircraft could be considered in-commission and they were only able to score 710 points out of a possible 1,000 – giving them a combat effectiveness evaluation of only 10%. The problems centered around   p73 the electrical and hydraulic systems on the aircraft, a never-ending problem for the ground crews. There always seemed to be a shortage of spare parts and the radar equipment had an average "lifespan" of only ten hours between failures. Cold weather also brought on a rash of landing gear failures, particularly in the tail wheels, as the ports in the hydraulic cylinders were too small to provide adequate pressure to extend the wheels when the fluid congealed.

A cold war flare-up occurred on October 3, 1949 and four F-82's were dispatched at Marks Field, where they flew a total of four missions over the next six days before returning to Ladd. An intelligence report of an impending Russian landing between Nome and Kotzebue was received through an undisclosed source and its validity had to be ascertained. The AF personnel assigned to Marks Field and the adjacent GCI site were given instructions on the methods to be followed if their installations had to be abandoned. "Walk out overland and don't associate with the natives." was the official "word." Many, if not all the Eskimos in that area were suspected of being communist sympathizers (since their heritage was strongly influenced by the earlier Russian possession of Alaska).

After the F-82's returned to Ladd, the GCI site was packed up and flown out, according to a control tower operator who was there at the time. In the morning it was functioning and, by mid-afternoon, it was gone. The west coast of Alaska was considered to be an indefensible position. If the Russians wanted it, it would be yielded and the U.S. would take steps to recover it later. A similar incident had occurred in 1948 when the 65th Fighter Squadron of the 57th Fighter Group was there with F-51H's and

they were pulled back to Anchorage when it was feared that the Russians might cross the Bering Strait in an effort to counterpoint the Berlin blockade.

When the constant adverse weather did permit flying operations during the winter of 1949-50, it was virtually a maximum effort to get in a maximum number of flying hours by the F-82 crews. It was found that it was unsafe to fly the F-82 when the surface temperature was below -20 F due to inadequate defroster heat. The canopies would frost over as soon as engines were started and they would never clear enough to provide visibility for safe flight. Cold soak also kept most of the other aircraft components from working properly.

When the Korean War started, the 449th, now redesignated as a Fighter (All-Weather) Squadron, was immediately placed on full combat alert status. Within a week they were indirectly supporting the war effort by conducting missions over communist territory. There were several Russian airfields and municipalities along the coastline of western Russia which had been built up during WWII which now deserved special attention – in case the communists were trying to build up their forces to support the North Korean effort. It would not have been politically expedient to use RB-29's or RB-36's to gather intelligence in this area, for, if one of them was downed, it would have been difficult to explain the violation of Russian airspace. Yet, if an F-82 or two went down, it could be passed off as a "navigational error" – or something of that sort. Hopefully, it would be treated by the Russians as a simple border violation.

The exercise was known as Special Mission Operations #122A. and 5 F-82H's departed Ladd Field for Marks Field at 2250 hours on July 1. 1950, arriving at Marks at 0100 hours   p73 on July 2. Two separate missions were planned, with the first to patrol the Russian coast adjacent to Avadyrskio Zalive (Yandrakinot) while the second was to patrol the Chukotskiy (Chukchi) Peninsula.

For these flights the aircraft's radar observers were issued hand held cameras. The first flight was airborne at 1516 BST but, due to low ceilings with rain and fog, the two F-82's returned within three hours without seeing anything. Nothing worth photographing was seen on the next two missions either.

On the fourth however, the aircrews were given a jolt when they spotted two returns on their radar scopes just south of Chaplina. Again, due to weather conditions, they could not get close enough to photograph anything.

strip. An RB-17, which the USAF decided to send along on this mission, also observed two additional groups of 10 Russian fighters each, believed to be La-11's. The communist aircraft were in loose gaggles instead of a true formation and they attempted

Sunrise at Galena Air Force Base, Alaska. A pair of 449th F(AW)S Twin Mustangs cast long shadows - Quartuccia

On July 10 another F-82 was scrambled against a target southwest of Marks Field and identified it as a B-29, so this F-82 was back on the ground within a half hour. A reconnaissance mission was then dispatched, but only two surface vessels were seen, one light and one heavy cargo ship. The missions continued through July 15 without spotting anything worthwhile. Then the detachment had to stand-down for a week because of weather and required maintenance on their aircraft. Later reconnaissance missions during the month of July were uneventful, but on August 1 everything went right for the mission to Avadyrskio. The visibility was perfect and six underground hangars and 12 C-47's were spotted and photographed. At Chaplina four Russian fighters were seen milling about at 1,000 feet over the air-

to come between the USAF aircraft and the Bering Strait. Due to the speed of the F-82 the La-11's were easily outdistanced while they had the Russian pilot's attention. After the RB-17 made its own retreat, the F-82's broke for home. However, with this threat of retaliatory air power, Mission #122A was curtailed.

Prior to the advent of the Korean War a shift in strategic planning had taken place which concentrated the 11th Air Force and its successor, the Alaskan Air Command, at Anchorage, in lieu of attempting to maintain fighter squadrons all across the Alaskan Theater. This was the move that saw the withdrawal of the 57th Fighter Group from Sheyma where they were flying F-51H's and the 449th FS(AW)S from Davis Field. Logistically, it was almost impossible to keep what were front line units supplied with spare parts and other required items. Being scattered across Alaska actually made them more vulnerable to loss than they would have been otherwise. The 449th FS(AW)'s detachment at Marks was again withdrawn into central Alaska after the mission #122A episode. Future operations would use Galena Air Force Station as a staging base. It remained a situation where, if the Russians should decide to attack, they would be almost welcome to it, for they would probably have as much trouble trying to hold on to what they captured as the Alaskan Air

F-82H
Radar Operator's Cockpit - Left Side

1. Anti G suit valve
2. Radar receiver circuit brkr.
3. AN/ART-13A transmitter circuit breaker
4. Anti G suit connection
5. SCR-720 ant, circuit breaker
6. Heat and vent control
7. SCR-720 Control panel
8. Auxiliary microphone sw.
9. Oxygen regulator
10. Cockpit air temp. control
11. Heater fuel pressure switch
12. Cockpit light rheostat
13. Circuit breakers
14. SCR-720 indicator tilting adjustment
15. Emergency air coolant flap release

F-82H
Radar Operator's Cockpit - Forward View

1. Clock
2. Oxygen pressure indicator
3. Oxygen flow indicator
4. Airspeed indicator
5. Altimeter
6. Free air temperature gage
7. BC-1151-B (SCR-720)
8. Defrost control
9. BC-1148D (SCR-720 synchronizer
10. Antenna hand control
11. Chartboard retainer
12. Radio compass indicator
13. Remote indicating compass

Command would have in trying to prevent such an attack.

At Galena AFS, the 449th F(AW)S maintained a detachment of 20 man and two F-82H's on constant alert. They were supported by 20 additional men of the Airways and Air Communications Service

F-82H
Radar Operator's Cockpit - Right Side

1. SCR-695B Radar control panel
2. AN/ARC-3 Radio control panel
3. Canopy handcrank
4. Command and interphone radio transfer switch
5. BC-453 and BC-454 radio control panel (F-82H Only)
6. AN/ART-13A transmitter control panel (F-82H Only)
7. Heat and vent outlet
8. Map case
9. Cockpit air volume control
10. Relief tube
11. Emergency canopy release
12. AN/ARC-6 Radio compass control panel
13. AN/APG-28 Antenna hand control (F-82F Only)

(AACS) who manned the control tower, Ground Controlled Approach (GCA) and the field's radio aids. In addition, there were 20 more men from the Air Base Unit (BASUT) who took care of the entire airstrip. As in the case of Marks Field, these men were all instructed on how to abandon the field and walk out overland to Fairbanks if they came under attack. There were no roads between Galena and anywhere else. If the strip could not be held for 24 hours by the 60 men, it was to be evacuated. "Use your own best judgment on whether or not it can be held." were their instructions.

Feint and challenge games with Russian aircraft continued. Russian reconnaissance aircraft would be spotted overflying parts of Alaska and they would be scrambled on by the F-82's of the 449th, now designated as the 449th Fighter Interceptor Squadron. It was a regular-as-clockwork operation. On any day after an observed overflight by a Russian aircraft, two F-82's would arrive at Galena prior to daybreak. They would then be refueled and their guns armed. This was followed by a takeoff and a terrain hugging departure while heading west, where they would overfly Russian bases and then return to Galena to refuel and have their guns safed before flying back to Ladd.

Attrition through accidents and airframe fatigue steadily reduced the number of F-82H's to an unacceptable number. Several of the ex-52nd and 325th F(AW)G F-82F's were winterized to F-82H standards. They were delivered to the 449th in 1951, but it was not too long before these aircraft also exceeded their own airframe life and they too had to be replaced. This left the USAF in something of a quandary as the F-94's in use in stateside interceptor squadrons had insufficient range to be of practical value as a tactical fighter over the expanse of Alaska. The solution was to bring to Alaska the remaining operational F-82G's which were in the inventory. These belonged to the Far East Air Force, serving in Japan and Korea in both air defense and ground support roles. The shipment of the 17 remaining F-82G's and four F-82F's, which had been flown in combat, began in May 1951 and continued through October. These aircraft were not the solution either, as they were all high time aircraft and most suffered from corrosion to some degree.

The 449th FIS, from mid 1952 through October 1952, utilized the F-94 for air defense, while the F-82's were used for ground support in tactical training exercises in conjunction with the U.S. Army. The last operational F-82, 46-377, was retired by the 449th FIS on November 12, 1953. Its passing marked more than the end of the Twin Mustang era as it was also the end of the conventional fighter in the USAF. (The Air National Guard, however, stayed with propeller driven aircraft for a few more years. The F-47 Thunderbolt was retired by the ANG in late 1954. The Air Force Reserve converted from F-51's to F-80's by 1953, and the ANG surrendered their last Mustang in 1957.) Two of the F-82's that were scrapped at Ladd Field were those which had scored kills in the opening dogfight of the Korean War, and everything considered, it was an inglorious end to an unusual aircraft.

Chapter Five

The F-82 proves its competency as a true combat aircraft...

In Korean Skies

The 347th Fighter Group (All-Weather) was reactivated on February 20, 1947 as a P-61 outfit. At the time, the Black Widow was considered an "old man's aircraft" – slow, but steady. Beginning in July of 1949, they started to receive all 45 examples of the F-82G. Deliveries were completed in August of that year. However, in somewhat of a historical anachronism, the 347th was deactivated on June 24, 1950 and all three of its squadrons were assigned to other fighter groups. The 4th Fighter Squadron (All-Weather), which had been a part of the original 52nd Fighter Group until their assignment to the 347th, was taken over by the 51st Fighter Wing at Naha, Okinawa with 13 of the new '82-G's. The 68th Fighter Squadron (All-Weather) was attached to the 8th Fighter-Bomber Wing at Itazuke, Kyushu, Japan with 12 of the new birds. The 339th FS (AW) ended up with the 35th Fighter Group. Their 14 F-82G's were divided between their main base (Johnson AFB, near Tokyo) and Misawa AFB on the island of Hokkaido. The two Japan-based squadrons were part of the 5th Air Force, while the F-82's of the 4th FS (AW) belonged to the 20th Air Force.

Since there were so few of the new Twin Mustangs in the Far East Air Forces (FEAF), there was much misgiving on the part of the "high priced help" as to whether or not the P-61 "drivers" could handle them. A call went out for Mustang pilots who could keep up with the new aircraft! FEAF polled all its P-51 outfits for volunteers who had lots of instrument time. There was some apprehension because the airplane looked so weird, but eventually almost all of the squadrons' rosters were made up of ex-Mustang pilots.

With the crews successfully transitioned into the Twin-Mustang, the F-82's were added to the "game" being conducted with the Russians. Before the Korean War began, the USAF ran RB-29 missions on a semi-regular basis to the Vladivostok area. One of these flights was jumped by Russian fighters who did enough damage to the reconnaissance Superfortress that the pilot had to crash land on Hokkaido Island. After this incident, the reconnaissance missions were made with fighter escorts, frequently the F-82G's. More usual was the use of F-80's to shepherd the recce aircraft but the missions were dropped entirely with the advent of the Korean conflict. Before that happened, however, other such "games" were played in reply to Russian feints toward Japan. One of the most popular included two flights of F-80's. One flight would fly at wave-top height toward Russian airfields. The second flight would fly high enough to be detected by Russian radar. Timed properly, the low flight would blast over the airfields just as the Russian fighters were scrambling to intercept the high flight. Then, everybody would streak for home at max speed! It was all "good clean fun."

These "games" had some bearing on F-82 operation in Japan prior to the opening of Korean hostilities, although the logic behind them seems questionable. As noted above, the 339th was based at Johnson AFB with four alert '82's at Misawa. Under the guise of conducting "maneuvers," the 339th was ordered to Itazuke to join with the 68th. They had instructions to leave four F-82's at Johnson and the detachment at Misawa. The rest of the Twin Mustangs flew south, much to the consternation of the squadron since splitting a squadron in this fashion was "never done." Everyone concerned was quite aware of the "game" being played up north and had difficulty understanding why the bulk of the squadron was being moved off in another direction.

The day the 347th FG (AW) was deactivated also marked the conclusion of the annual Far East Air Forces (FEAF) gunnery meet. All of the participating crews (including the 68th and 339th FS (AW)'s) returned to their home bases. This gave them just enough time to unpack their gear prior to being alerted of the Korean invasion on June 25, 1950.

Four of the 339th FS (AW) F-82G's at Misawa AFB and the main portion of the squadron at Johnson AFB were ordered to Itazuke AFB to join, once

Loading 5" HVAR's on an F-82 in Japan, August 1950. - Frazer via Air Force Museum

Maintenance on a 68th F(AW)S F-82G, August 1950. All of the plumbing shown gives an indication of just how complicated the Allison V-1750-143/145 engines were. - Frazer via Air Force Museum

again, with the 68th FS(AW). Also ordered to Japan was the 4th FS(AW) from Naha AFB and, by the evening of June 25, they had arrived with ten F-82's. Thus there were 34 Twin Mustangs at Itazuke, ready for action. This reunion of the three F-82G squadrons also brought about the temporary reactivation of the 347th FG(AW), under the command of Lt. Col. Jack Sharp, who was the highest ranking officer in the three squadrons. The 347th FG(AW) was placed under the direct control of the 8th Fighter Wing, which had the responsibility for tactical airpower in the 5th Air Force.

Early on the morning of the 26th June, FEAF received word from General MacArthur that he wanted fighter cover provided for the evacuation of American dependents from Korea. General Jarred V. Crabb, FEAF Director of Operations, passed the

The snow-swept flight line at Misawa AFB, Japan. Note that the two SB-17's in the background both mount their .50 caliber tail guns. One of the tailgunners came close to shooting down an Australian Mustang during the opening weeks of the Korean War. - White

A bomb is pulled into position for loading on an F-82. The USAF made heavy use of indigenous labor in Japan, due to personnel shortages. - Fraxer via Air Force Museum

"word" to Colonel John M. Price, Commanding Officer of the 8th Fighter Wing. Price advised General Crabb that his F-80's didn't have the necessary range to provide extended coverage over Inchon, the evacuation port. He suggested that the long-legged F-82's be tasked with the mission. The use of the Twin Mustang was approved and all F-82's in the Far East were gathered at Itazuke.

The first combat mission for the F-82's was to perform patrol duties and a top cover escort for the Norwegian ships *Reinholte* and *Norge* which were evacuating United Nations and U.S. dependents from South Korea. They also covered truck convoys shuttling between Seoul, Suwon and Inchon. The '82 crews weren't particularly impressed by the speed of the evacuation operation. Constant patrolling was hard, tedious work and most of it was under a 300 to 500-foot ceilings. Rain and haze made flying even more stressful.

Things got a bunch more dicey on the afternoon of the 26th. The 68th "had the duty" and the flight

ond La-7 started to fire at the second F-82 flown by Lt. George Deans, but his aim was also bad. Apparently, the communist fliers had seen enough and they disappeared in the mist. Well they did because Hudson and Deans had decided that they would return fire if they were attacked again.

Deans and Hudson climbed up through the clouds to join the remainder of their flight. The other two

Who says crew chiefs aren't athletic? Note the sway brace "kicker" mounted to the aft end of the drop tank. These were used to force the tank away from the wing in the event that the tanks had to be dropped, but they were only 50% effective in preventing damage to the wing. - Davis

had been orbiting Inchon harbor for a half-hour when they were diverted to Kimpo. Their task was to cover a convoy moving from Seoul to Inchon. As in all aerial combat, things suddenly got a lot more "interesting." A pair of Lavochkin La-7's showed up, coming from the direction of Suwon!

The first F-82, flown by 1st Lt. William "Skeeter" Hudson, with Lt. Carl Frazer as his radar observer, spotted the La-7's as they swung in behind them. Hudson ordered the flight to go "Gate" (maximum power), turn on their gun switches and drop their external fuel tanks. The lead La-7 started firing, but Hudson accelerated out of range while the sec-

F-82's were flown by Lt. Robert Bobo and Lt. Charles Moran. Moran just happened to be flying Bobo's personal aircraft, *B.O. Plenty*. When the order to drop the external fuel tanks had been given by Hudson, Moran's starboard tank came off in such a fashion as to tear off his pitot tube and damage his right aileron. Moran elected to land at Suwon, Korea to have it repaired. A C-54 was dispatched from Japan with the necessary parts along with Sergeant James Mehring, who was to effect the repairs. Before the damage could be repaired, a pair of Yak's attacked the airfield. Two 500-pounders destroyed both the C-54 and the F-82. Sgt. Mehring caught a bit of bomb-blasted concrete in his posterior. The airstrip was soon overrun by North Korean troops and it was the last anyone saw of the crumpled F-82.

The problems caused by dropping the external fuel tanks from F-82's was an ongoing one. Quite

David H. Trexler, R/O Lt. Victor Helfenbein. Number three was Lt. Walter Hayhurst, R/O Lt. Cliff Mills. The fourth F-82 aborted because its pilot got sick. The 4th FS(AW) flew top cover position at 12,000 feet and did not join the ensuing combat.

Capt. David H. Trexler in his office. Trexler was credited with a "probable" La-7 on June 27, 1950 - Trexler vis Olmstead

often the tanks would "roll off" their pylons instead of falling free. Damage to ailerons, pitot tubes and radio antennas was the usual result. Ultimately, a "fix" was devised – sway-brace "kickers" that would force the tanks away from the wings. Unhappily, they were only about 50% effective.

To cover further evacuation of "non-essential" allied personnel from Korea, the three squadrons of the 347th FG(AW) each dispatched a flight of four F-82's on June 27. The low flight, provided by the 68th, was composed of Lt. Hudson and his R/O Carl Frazer, Lt. Gene Planchek, R/O unknown, Lt. Robert Bobo, R/O unknown and Lt. Charles "Chalky" Moran and R/O Lt. Fred Larkins. The 339th FS(AW) flight drew the medium altitude and was led by Major James "Poke" Little, the squadron commander. Although the identity of his R/O is unknown, he is believed to have been Capt. Phillip B. Porter, the first R/O to achieve Ace status during WWII. The number two aircraft was flown by Capt.

The mission, beginning at 0930, Korean time, was led by Major Little and they were in a clockwise orbit over Kimpo Airfield at 8,000 feet, while the 68th FS(AW) orbited counterclockwise at 4,000 feet. The 4th's flight was out of sight above a thin overcast. Most of the time was spent just watching the C-47's and C-54's loading and departing.

At about noon, after two hours and 45 minutes of boring patrol between Kimpo and Suwon under lowering clouds that forced the aircraft as low as 2,500 feet, Lt. Hayhurst recalled someone exclaiming, "They're shooting at me!" Major Little gave the order to "break" and Hayhurst saw several birds that look similar to T-6's. Apparently it was Moran who had made the call alerting the flights, for his F-82 was the only one damaged by enemy fire. He received hits in his tail section.

Hayhurst swung his '82 in behind one of the hostile aircraft and started firing. Both he and his R/O (Mills) observed hits along the entire aircraft for about three seconds. He closed to 100 feet behind the Yak and was forced to break-off the attack. As he did, he almost hit another '82, head-on.

Moran, hit by machine gun fire, pulled into a vertical climb, to a point where his plane stalled and

onds before he had to break off the attack to avoid hitting another F-82.

At this time Major Little was heard to exclaim that he had just shot down an La-7. Hayhurst was bearing down on another La-7, but had to break off

Debriefing after the first air combat of the Korean War. (L to R) Capt. Johnie Gosnell, Lt, Charles "Chalky" Moran, Unidentified Sergeant, 1st. Lt. William "Skeeter" Hudson and Lt. Carl Frazer. - Frazer

this allowed Lt. Hudson to close behind his attacker, a Yak-11, who then tried to evade by climbing into the clouds. Hudson stayed with him and blasted the Yak with .50 caliber fire. The Yak pilot was seen to look and ascertain the condition of his observer who was apparently, dead. He then baled out. He landed near a group of South Korean soldiers and was dispatched after drawing his pistol.

As Moran recovered from his stall, he found himself on the tail of an La-7. He fired at close range and saw it dive straight into the ground. Hayhurst was behind another North Korean fighter. Both he and his R/O observed hits from their machine gun fire striking the aircraft for approximately three sec-

because he was just one of three F-82's chasing it. The La-7 broke right into Lt. Trexler's line of fire, but he was out of range. The La-7 turned again and Trexler cut him off. After scoring several hits on the communist fighter, Trexler observed it dive, inverted, into a cloud bank. He broke off the attack because the mountain tops were poking through the clouds indicating that it would not be a safe place for a descent. Although Texler didn't see the "hostile" crash, his R/O, Helfenbein verified that it had been hit several times. Evacuees saw an airplane plunge into the ground, but there were was no way of determining which pilot had shot it down. Both Trexler and Hayhurst were credited with probables.

Exactly which pilot had made the first Korean War kill will never be known. In fact, it took the

Two other factors cloud the "first kill" issue. During interrogation following the combat, Lt. Hud-

Carl Frazer briefs his wife, Pat, on the day's experiences, June 27, 1950 while "Skeeter" Hudson waits his turn on the phone.
- Frazer

USAF until 1953 to declare that Lt. Hudson was the first. There was a great degree of conflicting testimony about who made the first kill. Lt. Robert Bobo was flying the mission as Moran's element leader when he saw the first enemy aircraft bearing down on his wingman. He saw Moran whip his F-82 around and shoot it down. Moran may have been the first to make an aerial "kill" in the Korean War, but he was a competent, self-effacing pilot. He made no such claim. Hudson felt that he was the first but he made no effort to push the matter, either. The Johnson AFB newspaper, "The Gunner" credited Major Little as being the first – which is something he never said of himself. The 68th beat the 339th back to Itazuke after the combat and, thus, got there first and received the "best press."

son claimed an Su-2, as it most closely resembled the aircraft he shot down. In all probability, his victim was a Yak-11 as the USAF reports indicate. The Yak and the Sukhoi were very similar in appearance. Yak's were known to be in the theater while no confirmed reports indicate the presence of Sukhoi types.

Just to add further confusion to an already complex situation the question of just what pilot was flying which aircraft. Most sources state that Lt. Hudson was flying F-82G, 46-383, which was later named *Bucket Of Bolts*. Airplane #383 was assigned to the 68th, yet it was not Hudson's personal aircraft. His bird, 46-376, was down for maintenance at the time of the combat. Carl Frazer took photographs of the damage to Charlie Moran's aircraft after they landed, which show it to be 46-357. The 68th's

Yearbook indicated that Lt. Moran was flying #383 and that IT was the first aircraft to shoot down an enemy aircraft during the War. Neither Hudson nor Frazer remember the serial number of the aircraft they were flying that day!

At any rate, for the first taste of combat, the F-82 distinguished itself and answered any question

Was the the first air combat victory in Korea scored in this aircraft? There may be no earthly way to tell!

which might have arisen as to how such an ungainly appearing aircraft could maneuver in combat.

There were no missions flown on June 28 due to adverse weather, but, on June 29, four 339th F-82's (mission call sign "Anthony") were scheduled to escort General MacArthur's C-54 (mission call sign "Cleopatra") to Suwon. One F-82 aborted because its pilot got sick (again) and another because of mechanical problems. Lt. Harry "Whiskey" White and his wingman picked up the C-54 over the coast of Korea and escorted it to Suwon. After it landed Lt. White was contacted by Lt. Shermak, of the 36th FIS, who was serving as a Forward Air Controller. They were directed to fly north of the town on reconnaissance and interdiction mission.

They spotted several Russian-built T-34 tanks and proceeded to attack them until an engine on Lt. Frank Lee's F-82 quit. White had to shepherd him back to Japan. If they hadn't been called off on this mission, they probably would have been in position over Suwon for what became the greatest F-51 shootout of the war when Mustangs shot down four enemy aircraft and damaged a fifth. As it was, the F-82

never again joined in aerial combat.

The Korean conflict "spilled over" to Japan on June 29th when GCI radar spotted what was thought to be potential enemy aircraft. A "Red Alert" was sounded at Itazuke and fourteen F-82's were scrambled off a blacked-out runway. No enemy aircraft were spotted by the R/O's and the Twin Mustangs returned to base.

The following day, the weather went from bad to worse and the F-82's were the only aircraft able to get off from Itazuke for targets in Korea. There was no sign of enemy air activity and, after hitting some interdiction targets, the four F-82's of the 339th headed for home. They returned to find a ceiling reported somewhere between zero and two hundred feet. Since they all were low on fuel, they couldn't be diverted to an alternate field. Ground Controlled Approach (GCA) was saturated at Itazuke, for they were also "working" several flights of F-80's that couldn't return to their home bases. The first three 82's were vectored to safe landings but one, being flown by 1st Lt. Darrell Sayer was having problems. Apparently Sayer was having difficulties with his J-8 gyro because GCA couldn't get him to the runway. Sayer made three missed approaches and, on the fourth, was instructed to turn right to align him with the runway. He turned in the wrong direction and crashed into some low hills. Both Sayer and his R/O, 1st Lt. Vernon A. Lindvig were killed in the crash.

At this point, the F-82 was bearing the brunt of

the Korean (air) War, as it was the only aircraft available that could travel the 310 air miles from Itazuke to the Han River around Seoul and have enough fuel to loiter until a worthwhile target could be found under the low lying clouds. On June 29 the F-82's started carrying napalm tanks, which marked another first for them. The list of other "firsts" gained by the F-82 included: the first UN aircraft to fire a shot over North Korea, the first to shoot down an enemy aircraft and the first to provide true air support to the beleaguered ground forces.

The F-82G was not exactly suited to the ground support role. They were the night fighter version of

stick in a single 24-hour period!

The F-82 was a pretty rugged bird. On July 2, 1950, 1st Lt. Garling Peters of the 68th was attacking some enemy installation in the Inchon area when he was hit by a 20mm round in the cockpit. The shell exploded when it hit the throttle quadrant and tore away all of the controls except for one throttle. Petters was wounded by fragments in the neck, arms, body and legs yet he managed to fly the damaged aircraft back to Japan on one engine and then make a perfect landing.

On the night of July 2-3 F-82 "drivers" scored yet another first – radar night interdiction. Permis-

1st Lt. Harry "Whiskey" White of the 339th F(AW)S at Johnson AFB, Japan circa 1950. This F-82 was named "The Beast of the Far East but its serial number is unknown. White piloted one of the F-82 escorts for General Douglas MacArthur's C-54 to Suwon on June 29, 1950. - White

the Twin Mustang and, as such, were fitted with the radar "dong" between the fuselages. It would have been better if the Korean '82's had been the escort version, equipped with the North American-developed eight-gun pod. As it was, '82 crews pressed on with their ground attacks under very fatiguing conditions. Pilots were putting in long hours in the cockpit. One rugged soul put in 15 hours at the control

sion for these missions came from Major General Edward Timberlake, Vice Commander of the 5th Air Force, after being suggested by the 68th's C.O., Lt. Col. Sharp. The feasibility of these missions was demonstrated by Sharp and his R/O Capt. Cecil Wills, when they departed Itazuke and headed for the bridges across the Han River at Seoul. Navigating by Wills' radar returns, they penetrated an overcast five miles north of Seoul and broke out right on target, a bridge spanning the Han. They strafed it from one end to the other and returned to Japan with their point made.

On July 6, 1950, the 339th FS(AW) was released from their combat commitment and returned to

hazardous work, for the North Koreans would often string cables across the valleys for aircraft to fly into. Sometimes these cables were lighted. From the air they looked just like a convoy with its lights on.

During the month of August, 1950 the 68th FS(AW), in company with B-26's of the 3rd Bomb

Major James "Poke" Little, Commanding Officer of the 339th F(AW)S and Capt. Phi
liips S. Porten, his normal radar observer, walk to their F-82 prior to a mission. Note the taxi light hanging down just to the left of the bomb rack. The configuration of the flame dampening exhaust stacks is clearly visible. - USAF via Olmstead

Johnson AFB to resume their air defense duties. The 4th FS(AW) was released to return to Naha. Okinawa on July 8 and at this time the 347th FG(AW) was deactivated once again. The 68th FS(AW) now became the sole F-82G squadron to fly combat missions, although they were augmented by both crews and aircraft from the other two squadrons on a TDY basis.

Due to the shortage of F-82's, it was decided they would be only assigned to fly night missions, primarily interdiction. The radar observer would adjust his radar for ground mapping, and then, navigate by radar returns, skirting mountain tops and whistling through valleys until a target was spotted. It was

Wing (Light), averaged 35 sorties per night, which was quite a lot of air traffic over Korea. Cooperative sorties, between the '82's and the B-26's of the 3rd Bomb Wing saw the bombers illuminating targets with flares. The bright lights, however, hampered the fighter pilots' night vision and had to be curtailed.

Ground support strikes resulted in some unusual damage to the Twin Mustangs, but then, the F-82 was an unusual aircraft in the first place. It was found that pilots, flying attacks against ground targets, were encountering enemy small arms fire – which was to be expected. But one of two things was likely to occur as a result. Either the radar pod would be damaged or it would be "slung off" the aircraft when the pilots pulled too many "G's" while evading enemy anti-aircraft fire (normal operational loads restricted the F-82G to six "G's"). In either case, the F-82 was rendered "blind" for all-weather operations. Since there was an extreme shortage of '82 parts, a request was made to fly the radarless aircraft on daylight operations. The request was approved. As of the end of October, 1950, replacement pods had not yet arrived from the United States.

After the second deactivation of the 347th, the 339th resumed its air defense duties as part of the 35th Fighter Group. As they had previously, they continued to provide a detachment of three F-82's, crews and maintenance teams at Misawa AFB. One of the 339th's more interesting duties was providing fighter escort for Russian Embassy aircraft that shuttled between Japan and Russian airspace. Inbound Russian transports were picked up, usually, over Sado Shima island and accompanied to a landing at Tokyo's Haneda (civilian) Airport. Outbound flights were escorted from Haneda to Sado Shima.

As noted above, the 4th returned to Naha, Okinawa but, like the 339th, continued to supply aircraft and crews to the 68th. Historical records of the 4th's Korean War participation are not very illuminating but the 4th "officially" flew 43 combat sorties during their stint at Itazuke. They would continue to fly their '82's until April of 1951, when they reequipped with F-94B's. The 4th didn't get their 94's until late in the same year when they, also, received some F-80's

To return to Twin Mustang's Korean War exploits, they took part in the first large FEAF assault against the North Korean Army. On July 10th, an F-80 penetrated an overcast over Pyongtaek and spotted a large concentration of enemy activity. All available F-80's, B-26's and F-82's hit the area and raked the town for two hours. They wiped out 117 trucks, 38 tanks, seven half-tracks and an uncounted number of troops.

The F-82's worked best while alone at night. Lt. B.J. Buckhout recounted one night mission under these circumstances. He spotted a North Korean train. The engineer heard him coming and "jettisoned" his string of box cars and took off down the track for the cover of a tunnel. As he sped toward safety, he vented steam as rapidly as possible to prevent a boiler explosion if he was attacked. Attacked he was for Buckhout well and truly "ventilated" the engine with .50 caliber slugs. He then strafed the string of box cars!

Night missions would continue for the duration of the Twin Mustang's combat life and three are worth relating.

On August 25th, Captain Hudson (promoted since his June 27th exploit) and Lt. Frazer joined with another F-82 flown by Lt. George Broughton, R/O M/Sgt. Milton Griffin, for a dusk takeoff. They flew up the Naktong River (which edged the Pusan Perimeter) at low altitude until they received an urgent call for air support from a hard-pressed infantry unit. The UN troops were pinned down by communist fire from Hill 409. The pilots had to work under a 1,000 foot ceiling and were having trouble finding and keeping the hill in sight while flying in and out of rain showers. The forward air controller (FAC) made them repeat their passes until he was sure that they had positively identified the hill, lest they fire on UN troops. They were finally able to zero in and commence dropping their 500 pound bombs, and then fire off the 20 HVAR rockets each aircraft carried and finished up 45 minutes later with strafing runs.

August 30th, 1950 was recorded as the most memorable mission for 1st Lt. Ronald Adams, Jr. He and his wingman, 1st Lt. Robert Bobo, called the JOC (Joint Operations Center) for a target and were turned over to an LT-6 "Mosquito" working near Ainban-ni. The first two targets the Mosquito pilot picked were trucks, which were promptly dispatched. The third target's identification was never actually ascertained. It may have been either a truck loaded with ammunition or an ammunition dump – it was hard to tell in the dark. Adams bore down on it with his .50 calibers and it exploded upon being hit by the first burst of fire. A mighty shower of debris was thrown into the night sky, directly into the Twin Mustang's path. Happily, Adams' F-82 suffered only some minor scratches and scorching of its gloss black paint.

But, as mentioned, this was hazardous work. Lt. Moran and his R/O Lt. Francis J. Mayer had set out on one of these missions on August 7th and were lost in terrain so rough that the wreckage of their aircraft was not discovered for 18 months.

On September 7th, 1st. Lt. George Davis Jr. and his R/O, 1st. Lt. Marvin Olsen were flying their *L'il*

weather, he mistakenly made several strafing passes over Taegu, which he assumed to be the enemy-held

B. J. Buckhout bent this F-82 in a GCA landing accident after missing the strip and setting down on an unfinished section of the runway. The wreck was deemed repairable, although that doesn't seem obvious from the view above!

Bambi on a night interdiction mission . They were called in by a "Mosquito," an LT-6, to take out an observation post just outside of the Walled City which had been directing fire against UN troops. With only three rockets left from their previous attacks, they swung in and took out the post in one pass. They then continued to strafe the target until relieved by F-80's in the morning.

Misfortune struck UN forces on September 10th, through two incidents involving pilots of the 68th. The first occurred when Lt. Bobo mistakenly bombed a bridge which was in the hands of UN troops. They were desperately trying to keep it open to traffic. Bobo's two 500 lb. bombs hit the bridge's pillars and dropped the center section into the river. General Partridge personally interrogated Bobo after the incident because he had difficulty understanding how a single F-82 could knock out a bridge while the rest of the 5th Air Force was having such difficulty destroying the bridges over the Han River at Seoul. Bobo was returned to flying status but his wingman was so shaken by the incident that he asked to be relieved.

The second incident involved the 68th's Squadron's commander, Alden E. West. Flying in poor

town of Kunchon. Casualties were not heavy but one of the buildings he hit happened to be 5th Air Force Headquarters in Korea! General Partridge was standing on the building's steps and witnessed both the strafing runs and U.S. anti-aircraft gunners attempting to drive the F-82 off. Sgt. Anthony Palmari, West's R/O, had his parachute pack holed by a .50 caliber round. General Partridge personally relieved West of his command. He was replaced by Captain James S. Alford, Jr. until Major Little of the 339th could assume command of the 68th.

The sad events of September 29th were exacerbated by the loss of 1st Lt. Billy D. Stanton and his R/O, Capt. Robert MacDonald. Their F-82 collided in mid-air with an F-80.

B.J. Buckhout was shooting a GCA (Ground Controlled Approach) into Suwon in November, 1950. The sky was clear but the visibility was near zero due to blowing dust. The GCA controller told him that he was over the end of the runway when Buckhout was, in reality, just over the runway's overrun although he couldn't see it. Lt. Ayers, the R/O, spotted what he thought was the runway and Buckhout set the '82 down. Big problem! The Twin Mustang touched down on an uncompleted portion of the runway. When they hit the lip of the completed portion of the runway, the landing gear and both propellers were sheared off! Neither man was injured and the '82 was deemed repairable (!) despite the damage. Unfortunately, as it was being moved across the field for repairs, it slid off the flatbed trailer and the wings were crushed.

The 68th's Twin Mustangs prime task now became to fly weather missions. The "weather sorties"

by providing three F-82's for strip alert each night at the North Korean capitol. During this period, one aircraft would fly airborne weather reconnaissance and interdiction sorties while the other two sat, awaiting scramble orders.

On December 2, 1950, the 68th started to receive some relief from these missions. Four F-82's and crews from the 4th started pulling detached duty from Naha on a rotation basis. The 82's also got

Battle damage to the windshield of "Chalky" Moran's F-82, sometime in July 1950, and not the result of aerial combat. On the left is R/O 1st Lt. Clifford Pratt, who was killed on December 7, 1950 when his F-82 crashed after icing up on takeoff. The crew chief is unidentified. In the cockpit is Lt. Moran. He and his R/O. Lt. Francis J. Mayer, were killed on August 7, 1950 when their F-82 struck a cable strung across a valley while on a night mission. - Frazer via Air Force Museum

would take off between 0100 and 0200 hours, Korean time, and fly predetermined routes over North Korea. Their GCI control was known as "*Dentist.*" The 68th averaged fifty night sorties and 400 hours of flying time per month during the fall and winter of 1950/1951.

After Pyongyang was captured in November, 1950, the 68th, redesignated as a Fighter (All-Weather) Squadron, resumed an air defense function

some relief with the assignment of some F-7F-3N's from Marine Squadron VMF (N)542.

When Pyongyang was lost in December, the F-82's were pulled back to Seoul; then to Japan and back to Seoul again when it was recaptured in 1951.

Three F-82G's were lost through accidents in December, which left only 29 remaining in FEAF as 1951 began. On December 3rd, Capt. Dallas L. Jacobs took off from Kimpo on a "one-time ferry mission" to Itazuke. The '82 couldn't be repaired locally. Between Po'hang and Pusan one engine quit and the other one started running rough. Jacobs attempted an emergency landing at Po'hang but skidded off the runway and collapsed the landing gear.

On December 7th, Capt. Warren Harding took off from Kimpo in heavy icing conditions. Ice accumulated on his Twin Mustang so fast that he was unable to climb at all. He struck a hill some seven miles from the airfield and both he and his R/O, 1st Lt. Clifford Pratt, were killed.

The 68th F(AW)S used the tactical radio call sign of *"Vicious"* while operating over Japan and Korea. In December, 1950, a *"Vicious Special"* flight was attempted at night – to provide an airborne forward air controller for B-26 operations behind enemy lines. The operation didn't come off too well because of the difficulty involved in joining-up the F-82's and B-26's in the dark when the 82's spotted a worthwhile target.

It wasn't any safer flying in Japan during the winter period, either. On December 31st Capt. James Vahney, 339th Squadron, stalled his '82 on final approach to Johnson AFB while returning from an air defense scramble. His aircraft exploded when it struck short of the runway but Vahney was able to crawl from the wreckage before it was entirely consumed by fire. He suffered second and third degree burns.

Combat missions or sorties totaled 566 by the end of 1950. By then F-82s had fired 1,579 rockets, dropped 103 bombs and fired 276,000 rounds of .50 caliber ammunition. In spite of the fact that the 82's were the first to drop napalm on North Korea, official records only show a total of six napalm tanks having been dropped by them.

Combat missions would continue in spite of the attrition. So far six F-82's had been lost in the Far East to combat related incidents, six to accidents, one had been cannibalized for parts, one lost through an unknown cause and two lost in accidents prior to the Korean War. As a result of the attrition, the 68th F(AW)S was down to an average of only five serviceable F-82's. Because of the shortage of F-82's, armed interdiction missions were curtailed and missions now consisted of combat air patrols at night and air defense scrambles out of Kimpo (Seoul). This situation continued until the 339th started turn-

ing their '82's over to the 68th as they began to transition into jets in June of 1951.

Three missions may be considered as representative of those flown by F-82's in January. In one instance 1st Lt. Fred Murphy and his R/O, Lt. James Tullis took off after dark on an armed reconnaissance mission. Flying through heavy icing conditions to reach the target area, they broke out of the clouds over enemy territory. They descended into a winding valley along the road to Chorwon. They spotted thirteen vehicles. They pulled up to 500 feet, just below the cloud bases and then started a firing pass. After cutting loose with 5" HVAR rockets, which hit a fuel truck that lit up the area like daylight, they worked over the column until the enemy got their wits together and began firing back. They evaded the anti-aircraft fire by the simple expedient of climbing back into the clouds.

A similar mission was flown by Capt. Donald O'Neil and R/O 2nd Lt. Johnlion. When "their" exploding fuel truck lit up the area, all of the trucks in the convoy were illuminated – which were promptly dispatched.

The third representative mission involved 1st Lt. Laurence E. Anctil and R/O, 1st Lt. Robert L. Greer on January 26th. These two airman were assigned to the 4th F(AW)S but were pulling detached service with the 68th. They were assigned to fly a CAP (Combat Air Patrol) over Pyongyang. They arrived in the targeted area and relieved the crew already on-station. When their relief arrived, they could not be contacted. Both were declared missing in action and their deaths were confirmed after the war had ended.

As of May 31, 1950 there had been 32 F-82G's in Japan. On June 26th, the first one was lost at Suwon. This single incident opened the question as to what attrition would do to theater stock of the all-weather fighter. As General Frank Everest explained, at that time, the USAF had a total of 168 F-82's in the entire inventory and this number included the all-weather 82F's and H's in Alaska and Washington state. These Twin Mustangs would not be available to FEAF for sometime to come since they provided air defense to the U.S. The F-82E's, which had been used by SAC had already been pulled from the inventory and were in the process of being scrapped. Any attempt to "rescue" those that hadn't been disposed of and to modify them to all-weather standards was considered unfeasible. In short, if the Fifth Air Force was to continue to use the F-82 for armed reconnaissance, it was forecast that they would deplete all available spare parts within sixty days. Another factor to be considered was that the F-82's represented the only counter-air interceptor fighter that FEAF had during periods of darkness and bad weather. This was true for the entire USAF since the P-61 Black Widows had been withdrawn from service and the F-94 and F-89 were just beginning to roll off the production lines. Air Force-wide, the versatility of the Twin Mustang couldn't be squandered.

General Otto P. Weyland, commander of FEAF, had to consider the nighttime defense of UN bases in Korea. There had been only a few night intruders so far, commonly known as "Bed-Check Charlies" but there was still the threat of larger raids by IL-10's and Po-2's. Beginning in October, 1950, Weyland had ordered that two F-82's be staged from Seoul (Kimpo) to provide for defense against such attacks. To comply with this order, the 68th dispatched two aircraft, either alone or as a pair, on night interdiction missions. On completion of their strikes, they would land at Kimpo for refueling and rearming. They would then be spotted just off the end of the runway with APU's (Auxiliary Power Units) plugged in with the crews in the cockpits – just in case "Charlie" came along. If the aircraft hadn't been scrambled after a suspected intruder during the night, they would be released just as dawn was breaking to make a weather reconnaissance flight along the Yalu River - to provide an accurate weather briefing for 5th Air Force "Intelligence" officers. Other F-82's, in the meantime, would be departing Itazuke to fly continuous CAP over Pyongyang and Kandong airfields to prevent communist night operations from these airstrips. No other USAF aircraft were permitted within a ten-mile radius of these strips. The '82 crews were instructed to shoot first and then question the identity of any aircraft that violated their airspace. These missions made for many long and tense nights for the involved aircrew but they kept the communists at bay during their preferred time of maximum activity – at night. The night CAP's did save considerable wear and tear on the aircraft which ground support missions would have inflicted on them.

The Twin Mustangs were released for a resumption of interdiction missions in the winter of 1951. One F-82 was lost on the resumed nocturnal strikes on March 14th when Capt. Julius Fluhr and his R/O, 2nd Lt. Frederick Milhaupt took off from Itazuke at 0205 hours. They contacted a GCI site known as "Mellow" who informed them that they were heading beyond the bomb line. With the exception of a single further contact between "Vicious Seven" (68th Squadron) and another F-82, it was the last that was ever heard from "Seven." The next day an F-82 drop tank and two parachutes were spotted on the ground south of Munsan and Vicious Seven's crew was carried as missing in action. Their deaths were confirmed after the war ended. Their loss was assumed to have been caused by collision with a North Korean wire, strung across a valley. It was not known for sure but cables may have also taken the lives of Capt. Arlid Neilson and his R/O, Capt. Ralph Mulhallen on May 26th. They departed K-2 airfield at 0538 hours and were never seen again.

The last combat loss of an F-82 occurred on July 3, 1951 under some rather unusual, if not fortuitous, circumstances as far as the crew was concerned. The pilot, Capt. James Johnson and his R/O, 1st Lt. Roderick Doll, were flying a weather recce mission when antiaircraft fire started a fire in their right wing. Johnson headed for the west Korean coast, hoping that the fire would "starve itself out" – which it did. Clearing the coast, Johnson turned south, heading for friendly territory. Even as he turned, the fire started burning again and continued until there was an explosion in the fuel tank. At that time, the '82 was about 90 miles northwest of Seoul and the wing was burning like the proverbial haystack. Discretion being the better part of valor, the crew elected to bail out over the water. Both men were rescued by separate South Korean fishing boats and were later recovered by ARS (Air Rescue Service) Grumman amphibious SA-16's.

The 339th F(AW)S, now known as the 339th Fighter-Interceptor Squadron, was the first FEAF F-82 outfit to start phasing out of the Twin Mustangs for jets. They first received F-80C's from the 41st FIS for transition training during April, 1951. They took delivery of their F-94B's during the summer and fall months. The 339th, however, continued to supply F-82 crews to the 68th FIS (so redesignated during April as well) including four crews in

July. These men did not have sufficient time remaining on their overseas tours to be profitably jet-qualified.

The 68th continued to fly combat over Korea for the duration, flying a mixed bag of F-82G's and some F-82F's that had arrived from the States as replacement aircraft. The 68th received their first F-94B's during August and started transition training even while maintaining a strip alert contingent at Kimpo. One '82 was lost during August at Kimpo when, in a moment of poetic irony, a Twin Mustang named *Da Quake*, struck an earth mover that was parked too close to the runway.

In December, '51, the 68th began stationing two of their new F-94's at Suwon for night and bad weather protection of UN forces. It was during this period that the Starfire was restricted to flight over UN-controlled airspace because of its classified radar equipment. Thus, F-82's remained as alert aircraft in case a target would show up on a GCI controller's scope that might indicate a threat along or north of the battle line. This duty was now being shared between F-7F's and F-4U-5N's provided by the Marines.

The 68th's '82's remained in Korea until March, 1952, when they were flown to Itazuke to be modified to F-82H standards. They were then flown to Alaska to serve with the 449th FIS at Ladd AFB, Fairbanks. Fittingly enough, the last operational F-82 with the 449th was one of the veterans of the Korean War.

Appendix A

Drawings and Three Views

Note: The permission of Model Airplane News to reproduce
the Willis L. Nye three-views here is gratefully acknowledged

North American F-82G Twin Mustang Cutaway Drawing Key

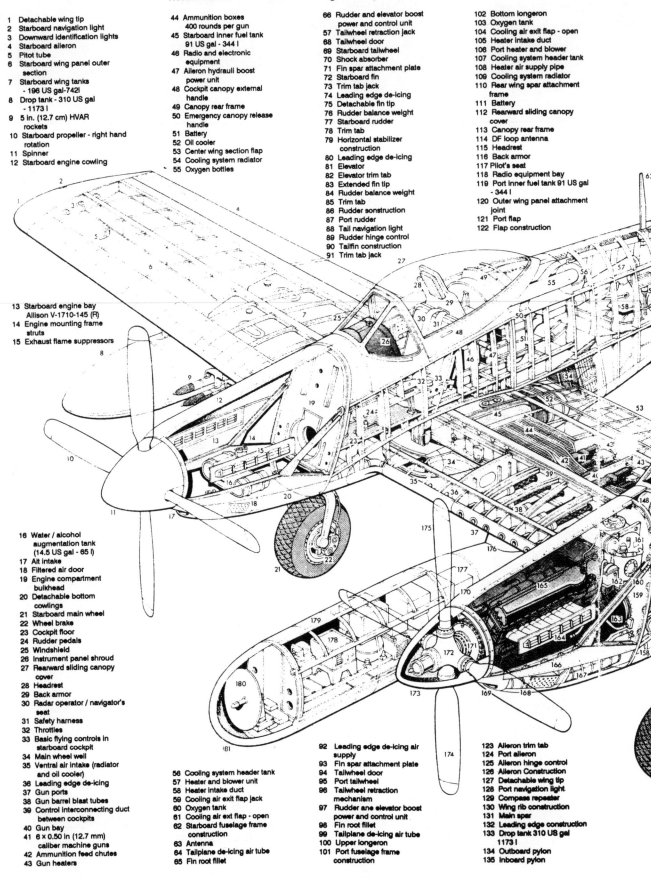

1 Detachable wing tip
2 Starboard navigation light
3 Downward identification lights
4 Starboard aileron
5 Pitot tube
6 Starboard wing panel outer section
7 Starboard wing tanks - 196 US gal-742l
8 Drop tank - 310 US gal - 1173 l
9 5 in. (12.7 cm) HVAR rockets
10 Starboard propeller - right hand rotation
11 Spinner
12 Starboard engine cowling

13 Starboard engine bay Allison V-1710-145 (R)
14 Engine mounting frame struts
15 Exhaust flame suppressors

16 Water / alcohol augmentation tank (14.5 US gal - 65 l)
17 Alt intake
18 Filtered air door
19 Engine compartment bulkhead
20 Detachable bottom cowlings
21 Starboard main wheel
22 Wheel brake
23 Cockpit floor
24 Rudder pedals
25 Windshield
26 Instrument panel shroud
27 Rearward sliding canopy cover
28 Headrest
29 Back armor
30 Radar operator / navigator's seat
31 Safety harness
32 Throttles
33 Basic flying controls in starboard cockpit
34 Main wheel well
35 Ventral air intake (radiator and oil cooler)
36 Leading edge de-icing
37 Gun ports
38 Gun barrel blast tubes
39 Control interconnecting duct between cockpits
40 Gun bay
41 6 x 0.50 in (12.7 mm) caliber machine guns
42 Ammunition feed chutes
43 Gun heaters

44 Ammunition boxes 400 rounds per gun
45 Starboard inner fuel tank 91 US gal - 344 l
46 Radio and electronic equipment
47 Aileron hydrauli boost power unit
48 Cockpit canopy external handle
49 Canopy rear frame
50 Emergency canopy release handle
51 Battery
52 Oil cooler
53 Center wing section flap
54 Cooling system radiator
55 Oxygen bottles

56 Cooling system header tank
57 Heater and blower unit
58 Heater intake duct
59 Cooling air exit flap jack
60 Oxygen tank
61 Cooling air ext flap - open
62 Starboard fuselage frame construction
63 Antenna
64 Tailplane de-icing air tube
65 Fin root fillet

66 Rudder and elevator boost power and control unit
57 Tailwheel retraction jack
68 Tailwheel door
69 Starboard tailwheel
70 Shock absorber
71 Fin spar attachment plate
72 Starboard fin
73 Trim tab jack
74 Leading edge de-icing
75 Detachable fin tip
76 Rudder balance weight
77 Starboard rudder
78 Trim tab
79 Horizontal stabilizer construction
80 Leading edge de-icing
81 Elevator
82 Elevator trim tab
83 Extended fin tip
84 Rudder balance weight
85 Trim tab
86 Rudder sonstruction
87 Port rudder
88 Tail navigation light
89 Rudder hinge control
90 Tailfin construction
91 Trim tab jack

92 Leading edge de-icing air supply
93 Fin spar attachment plate
94 Tailwheel door
95 Port tailwheel
96 Tailwheel retraction mechanism
97 Rudder ane elevator boost power and control unit
98 Fin root fillet
99 Tailplane de-icing air tube
100 Upper longeron
101 Port fuselage frame construction

102 Bottom longeron
103 Oxygen tank
104 Cooling air exit flap - open
105 Heater intake duct
106 Port heater and blower
107 Cooling system header tank
108 Heater air supply pipe
109 Cooling system radiator
110 Rear wing spar attachment frame
111 Battery
112 Rearward sliding canopy cover
113 Canopy rear frame
114 DF loop antenna
115 Headrest
116 Back armor
117 Pilot's seat
118 Radio equipment bay
119 Port inner fuel tank 91 US gal - 344 l
120 Outer wing panel attachment joint
121 Port flap
122 Flap construction

123 Aileron trim tab
124 Port aileron
125 Aileron hinge control
126 Aileron Construction
127 Detachable wing tip
128 Port navigation light
129 Compass repeater
130 Wing rib construction
131 Main spar
132 Leading edge construction
133 Drop tank 310 US gal 1173 l
134 Outboard pylon
135 Inboard pylon

48

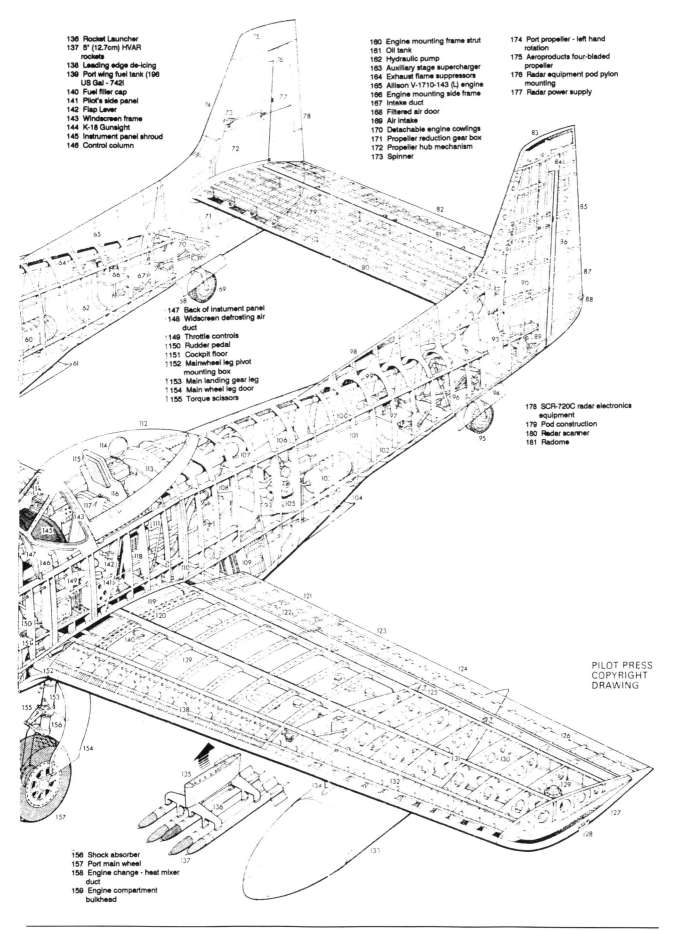

136 Rocket Launcher
137 5" (12.7cm) HVAR rockets
138 Leading edge de-icing
139 Port wing fuel tank (196 US Gal - 742l)
140 Fuel filler cap
141 Pilot's side panel
142 Flap Lever
143 Windscreen frame
144 K-18 Gunsight
145 Instrument panel shroud
146 Control column

147 Back of instument panel
148 Widscreen defrosting air duct
149 Throttle controls
150 Rudder pedal
151 Cockpit floor
152 Mainwheel leg pivot mounting box
153 Main landing gear leg
154 Main wheel leg door
155 Torque scissors

156 Shock absorber
157 Port main wheel
158 Engine change - heat mixer duct
159 Engine compartment bulkhead

160 Engine mounting frame strut
161 Oil tank
162 Hydraulic pump
163 Auxiliary stage supercharger
164 Exhaust flame suppressors
165 Allison V-1710-143 (L) engine
166 Engine mounting side frame
167 Intake duct
168 Filtered air door
169 Air intake
170 Detachable engine cowlings
171 Propeller reduction gear box
172 Propeller hub mechanism
173 Spinner

174 Port propeller - left hand rotation
175 Aeroproducts four-bladed propeller
176 Radar equipment pod pylon mounting
177 Radar power supply

178 SCR-720C radar electronics equipment
179 Pod construction
180 Radar scanner
181 Radome

PILOT PRESS
COPYRIGHT
DRAWING

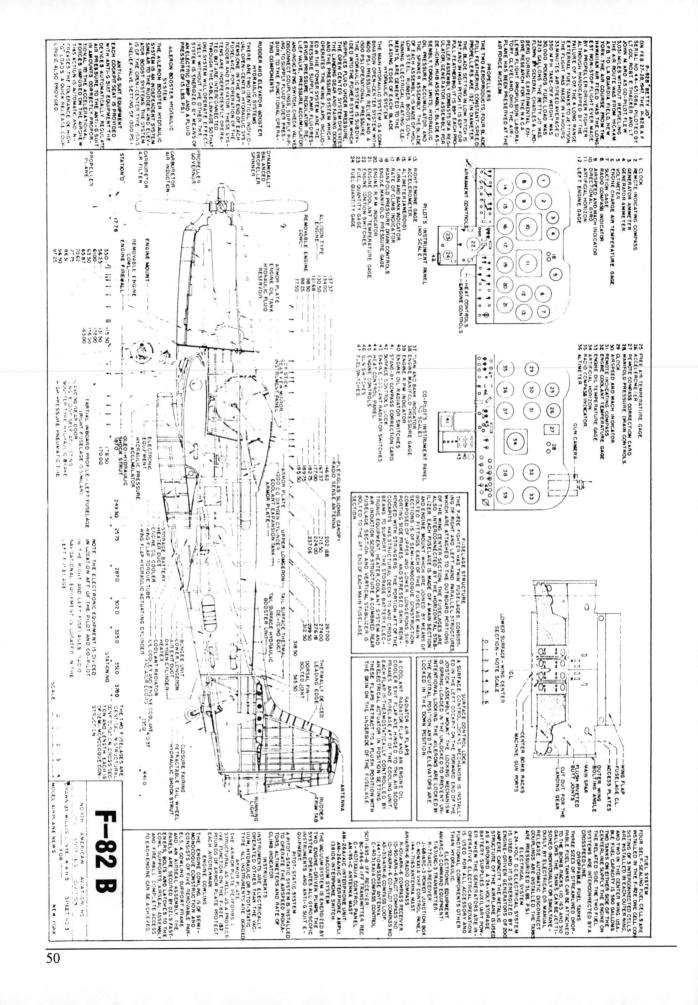

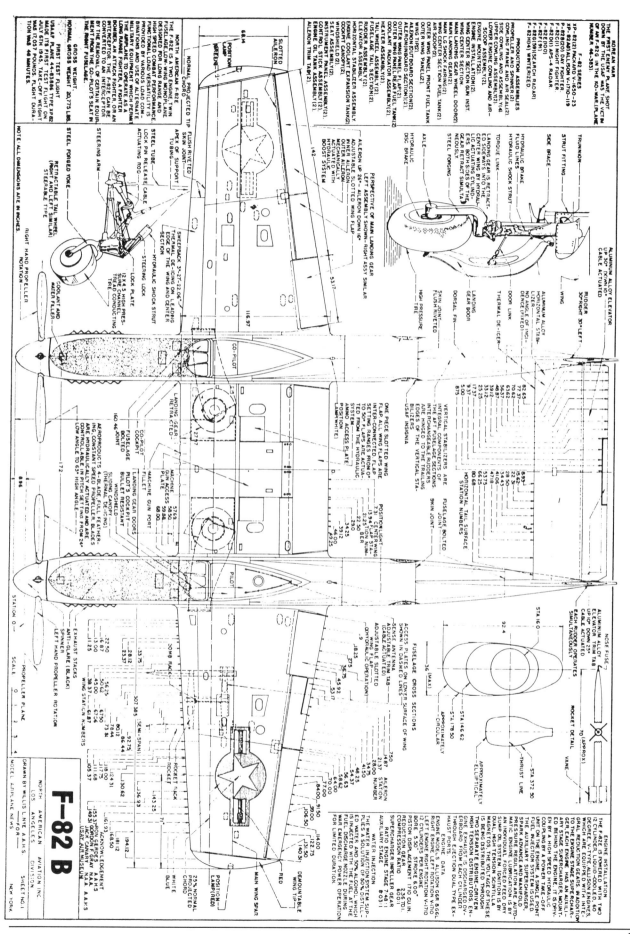

F-82 B

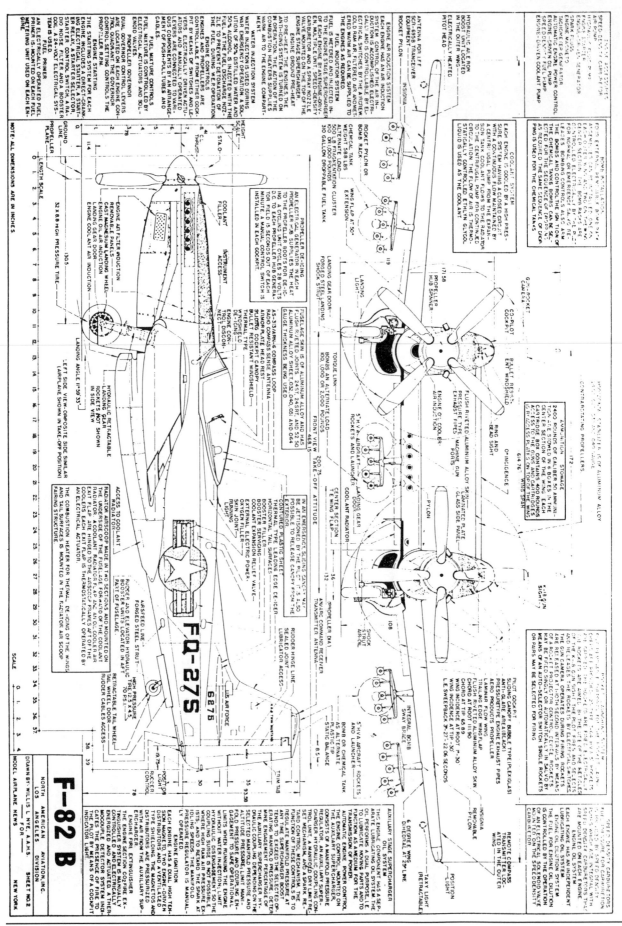

F-82 B

NORTH AMERICAN AVIATION, INC.
LOS ANGELES DIVISION
DRAWN BY: WILLIS L. NYE, A.A.H.S.   SHEET NO.2
MODEL AIRPLANE NEWS   NEW YORK

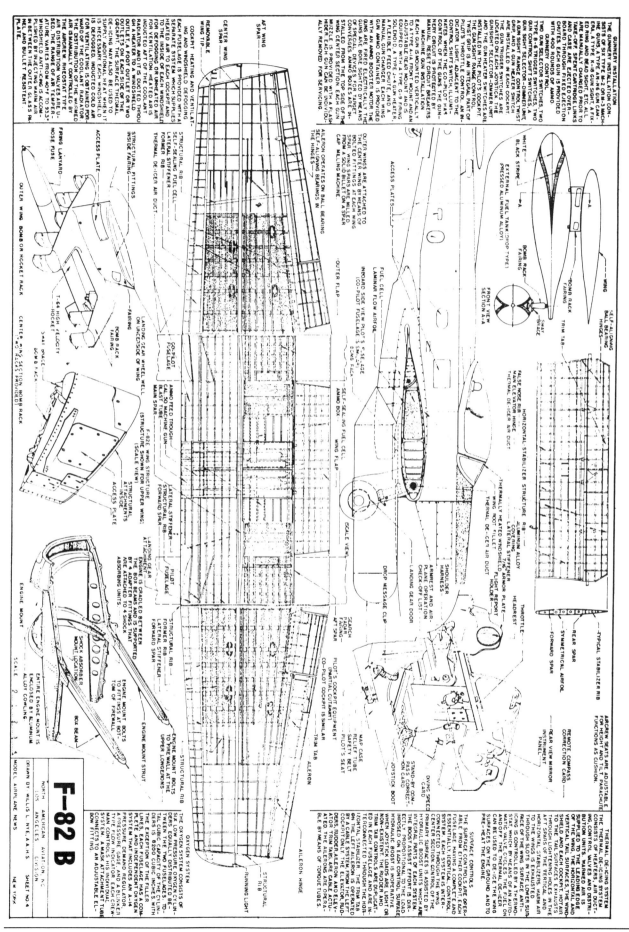

F-82 B

NORTH AMERICAN AVIATION, INC.
LOS ANGELES DIVISION

DRAWN BY WILLIS L. NYE, A.A.H.S.

MODEL AIRPLANE NEWS   NEW YORK

SHEET NO. 4

APPENDIX B

Supplemental Airframe Data
Oil capacity

37 gallons in two tanks located in engine accessory sections. Grade 1100. Oil coolant provided by separate radiators located in front of the engine coolant radiators in each air scoop. (These radiators also provided cooling for the hydraulic coupling oil for the auxiliary stage supercharger.) Airflow through the radiators is provided by the movement of thermostatically controlled outlet flaps.

Augmentation

A 14-25 gallon water alcohol tank is provided for each engine, located behind the propeller on top of the engine nacelle. Both engine systems were controlled simultaneously, and capacity provided for approximately ten minutes of operation. (Methanol was added to prevent it from freezing at high altitudes or at extremely cold surface temperatures. Anti-Detonant Injection (ADI) is the correct term for this feature, as it kept combustion temperatures at a safe level for air/fuel mixtures during takeoffs or under war emergency conditions.

Fuel System

Four self-sealing internal fuel tanks, consisting of two interconnected cells in each outer wing panel and two in the wing center section. The interconnected tanks held 200 gallons each, while the inner tanks each held 90 gallons. Total usable fuel was approximately 575 gallons. A cross-feed was available, although each fuel system was normally used for its own engine.

External fuel could be provided by a 110, 165 or 310 drop tank mounted on pylons under each outer wing panel.

Electrical System

Two engine driven generators supplied power a 28 volt DC system. A 24 volt, 34 ampere hour battery served as backup. Magnetos were manufactured by Delco-Remy, with two per engine.

Armament

Six Browning MG-53-2 machine guns mounted in the wing center section. 400 rounds per gun, K-18 reflector gunsight. AM-16 gun camera. J-4 and/or J-6 gun heaters.

Maximum bomb load 2 X 2,000 pound bombs. Maximum bomb size 2,000 pounds. 25 High Velocity Aircraft Rockets. HVAR (F-82E)

F-82F, G & H identical extent maximum bomb load of 2 X 1,000 pound bombs, maximum bomb size 1,000 pounds. 20 HVAR's.

Note: The XP-82 was rated to a total of 6,000 pounds of bombs, with an AN-M-65 1,000 pound bomb on each outer pylon and an AN-M- 66 2,000 pound bomb on each inner pylon. Alternate ordnance of an AN-M-65 on each outer pylon and an AN-MK-13 1,600 pound bomb on inner pylons.

Airfoil

Wing area 417.6 square feet. M.A.C. 100.8. Wing section NACA 66, 2-215 at root, NACA 66 1-212 at tip. Aspect ratio 6.28. Span 51' 2". Incidence 1 1/2° at root, -1.2^o at tip. Dihedral 6°. Sweepback 3° 27' (For F-82E).

Flaps had 45° travel. Chord 26.6% 49.5 square feet of area.

Ailerons were sealed balance type, with a total area of 12.35 square feet. 15° up travel, 15° down travel. Trim tab area 1.92 square feet.

Stabilizer maximum chord 4.83 feet. Span 14.33 feet. Area 52.8 square feet. Elevator span 14.3 feet. Total area 13.2 square feet, 25° up travel, 10° down travel.

Fin area 43.4 square feet. Rudders 15.2 square feet. 30° degrees travel in both directions.

Landing Gear

High pressure type tires with Channel tread 32" X 8". The wheels are Channel tread 12½ X 4½." Retracting system for both main and tail wheels is hydraulic. Disk type brakes.

Avionics

F-82E

AN/ARC-3 VHF Command
AM-26A-A1C Interphone
AN/ARA-8 Homing Adapter
AN/ARN-6 Radio Compass
SCR-695B IFF

F-82F

As above, with:
AN/ARN-5A and
RC-103A for instrument landings (provides vertical and lateral glide path information)
RC-193 Marker Beacon

AN/APG-28 Radar Search and Tracking.

F-82G

As above, excepting:
SCR-720C replaced the AN/APG-28.

F-82H

As F-82F excepting additional BC-453 and BC-454 for reception of low and high frequency radio signals
Also included:
AN/ART-13A long range liaison transmitter.

F-82 Serial Numbers

| NA-120 | XP-82 (2) | 44-83886 - 83887C/N 120-43742 -43743 |
| | XP-82A (2) | 44-83888 - 83889C/N 120-43744 -43743 |

There is no evidence that the XP-82A's were ever flown, due to problems in developing the Allison V-1710-119 engines. The second example was definitely terminated before construction was completed, and the disposition of the first never verified.

NA-123	P-82B (19)	44-65160 -65168C/N 123-43746/43754
	P-82C (1)	44-65169C/N 123-43755
	P-82D (1)	44-65170C/N 123-43756
NA-144	P-82E (100)	46-255 - 354C/N144-38141 - 38240
NA-149	P-82F (100)	46-405 - 504C/N 149-38291 - 38390
NA-150	P-82G (28)	46-355 - 383C/N 150-38241 - 38269
	(15)	46-389 - 404C/N 150-38275 - 38290
	P-82H (14)	46-384 - 388C/N 150-38270 - 38274
		46-496 - 504C/N 150-38382 - 38390

Appendix C

USAF P-82's utilized in the Korean War

The capital letters following the aircraft serial number denotes the cause of loss, as noted on the individual aircraft's record card. Quotation marks indicate the name of the aircraft. Parenthesis indicate the names of the crew flying when the aircraft was lost, or its normal crew.

A -Accident.
B -War Weary.
C -Crashed in Water.
J -Operational accident on a combat mission, or salvage if peacetime. Cannibalized for spares.
L -Lend Lease.
M -Enemy action on a combat mission.
N -Operational loss on a combat mission.
P -Unknown reason for loss on a combat mission.
R -Abandoned, due to enemy action.
Q -Corrosion.

F-82G

46-355 M 68th F(AW)S. "M" 8-7-1950 (Moran/Mayer)

46-356 A 4th F(AW)S, Write-off at Naha 8-10-1951. Ex-68th and 339th F(AW)S "Gluckin-Blackin" (Harp/Allen)

46-357 P 68th F(AW)S. "P" 5-26-1951. (Neilson/Mulhallen)

46-358 N F(AW)S "N" 6-30-1950 (Sayer/Lindvig)

46-359A 4th F(AW)S. "A" at Naha 11-7-1950 "Misguided Virgin" (Tymonicz/Jennings)

46-360 ? 4th F(AW)S. Write-off 9-28-1950 at Naha

46-361 68th F(AW)S. Received from 4th F(AW)S 12-19-1950. Returned to 4th F(AW)S 4-1951. To 449th F(AW)S, Alaska 5-1952 "Double Trouble"

46-362 339th F(AW)S. To 4th F(AW)S 6-13-1951. To 449th F(AW)S 6-1-1952

46-363 68th F(AW)S. Aircraft was bombed at Suwon. "B.O. Plenty" (Moran's)

46-365 339th F(AW)S. To 4th F(AW)S 6-29-1951. To 449th F(AW)S 6-19-1952

46-366 4th F(AW)S. To 449th F(AW)S 5-1952

46-367 A 339th F(AW)S. Write-off 3-2-1951 at Johnson. ("Lover Boy")

46-368 A 339th F(AW)S. Write-off 5-7-1950 at Johnson

46-369 J 68th F(AW)S. Write-off 5-11-1950 as cannibalized for spare parts.

46-370 J 13th Air Base Group. Write-off 2-31-1950. Cannibalized for spare parts.

46-371 N 68th F(AW)S. 8-12-1951 at Kimpo. Hit bulldozer. "da Quake"

46-372 339th F(AW)S. To 68th F(AW)S 7-23-1951. To 449th F(AW)S 5-20-1952

46-373 A 68th F(AW)S. 2-12-1951 (Boughton)

46-374 A 68th F(AW)S. 5-1-1951

46-375 N 68th F(AW)S. "A" 12-3-1950 at P'ohang (Jacobs)

46-376 A 68th F(AW)S. 9-28-1950 in mid-air with F-80. "Stormy" (Stanton/MacDonald)

46-377 339th F(AW)S. To 4th F(AW)S 6-19-1951. To 449th F(AW)S 5-20-1952

46-378 N 68th F(AW)S. 7-3-1951 (Johnson/Dall) Received from 339th F(AW)S 5-1951

46-379 339th F(AW)S. To 68th F(AW)S 7-9-1951. To 449th F(AW)S 5-20-1952 "The Dull Tool"

46-380 A 339th F(AW)S. Write-off 5-31-1950

46-381A 339th F(AW)S. Write-off 7-21-1950

46-382 4th F(AW)S. To 449th F(AW)S 5-20-1952 "Night Takeoff"

46-383 68th F(AW)S. To 449th F(AW)S 5-20-1952. (First 68th F(AW)S to pass the 1,000 flying hours mark. Hudson's aircraft when becoming the first USAF pilot to claim a kill. "Bucket of Bolts"

46-384 through 46-388 were F-82H's

46-389 4th F(AW)S. To 449th F(AW)S 5-2-1952 ("Wee Pee")

46-390 4th F(AW)S. To 449th F(AW)S 5-20-1952 ("Midnight Sinner")

46-391A 68th F(AW)S. Write-off 9-10-1950 "Li'l Bambi"

46-392 N 68th F(AW)S. From 339th F(AW)S 7-9-1951. 11-6-1951 at Suwon. ("Our Little Lass")

46-393 68th F(AW)S. To 449th F(AW)S 5-20-1952

46-394 P 68th F(AW)S. 3-4-1951 (Fluhr/Milhaupt) Received from 4th F(AW)S 12-3-1950 ("Dottie Mae")

46-395 339th F(AW)S. To 4th F(AW)S 6-24-1951. To 449th F(AW)S 6-20-1952

46-396 J 68th F(AW)S. Write-off 7-10-1950. Loss code changed from "M" to "J" 1-13-1951

46-397 4th F(AW)S. To 68th F(AW)S 12-5-1950. To 4th F(AW)S 4-25-1951. To 449th F(AW)S 5-20-1952

46-398 A 339th F(AW)S. Write-off 12-31-1951 "Black Jack" (Vahet)

46-399 P 68th F(AW)S. Write-off 1-26-1951 (Ancil/Greer) "Able Mabel"

46-400 A 68th F(AW)S. Write-off 12-7-1950 (Harding/Pratt) From 4th F(AW)S 12-3-1950 "Call Girl/Night Mare"

46-401 339th F(AW)S. To68th F(AW)S 7-9-1951. To 449th F(AW)S 5-20-1952. ("Gruesome Twosome") also ("Patches on the Patches")

46-402 M 4th F(AW)S. Write-off 7-6-1950 at Naha as result of battle damage

46-403 4th F(AW)S. To 449th F(AW)S 5-20-1952 "Miss Carriage"

46-404 339th F(AW)S. To 68th F(AW)S 7-9-51. To 449th F(AW)S 5-20-1952

F-82F

46-415 68th F(AW)S. From 52nd F(AW)G 5-10-1951. To 449th F(AW)S 5-20-1952

46-448 68th F(AW)S. From 325th F(AW)G 7-27-1951. To 449th F(AW)S 5-20-1950

46-473 68th F(AW)S. From Wright Patterson 10-28-1951. To 449th F(AW)S 5-20-1952

46-491 68th F(AW)S. From Eglin 7-27-1951. To 449th F(AW)S 5-20-1952

F-82's assigned to operational units.

27th Fighter Escort Group

F-82B

44-65126 27th FEG 1-20-1948. "HH" (Excess to requirements, reclaimed for parts.) 7-31-1948. 17 Hrs TT,

44-65173 27th FEG 3-2-1948 3-2-1948. To 325th F(AW)G 6-2-1948

44-65177 27th FEG ? To 57th F(AW)G 6-28-1948

F-82E

46-255 27th FEG 12-13-1948 "J" Robbins AFB 7-31-1950 432 Hrs TT

46-259 27th FEG 3-18-1949 "J" Robbins AFB 8-4-1950 458 Hrs TT

46-261 27th FEG 4-8-1949 "J" Robbins AFB 9-25-1950 604 Hrs TT

46-263 27th FEG 8-11-1949 "J" Robbins AFB 8-9-1950 447 Hrs TT

46-254 27th FEG 2-6-1949 "J" Robbins AFB 8-4-1950 395 Hrs TT

46-266 27th FEG 5-16-1948 Write-off Kearney AFB No Code 3-8-1949 134 Hrs TT

46-267 27th FEG 5-11-1948 "J" Robbins AFB 8-4-1950 567 Hrs TT

46-268 27th FEG 6-1-1948 "J" Robbins AFB 2-11-1950 410 Hrs TT

46-269 27th FEG 5-20-1948 Write-off Hill AFB No Code 12-16-1948 No flying hours listed

46-270 27th FEG 6-1-1948 "J" McClellan AFB 4-28-1950 510 Hrs TT

46-271 27th FEG 6-11-1948 "J" Robbins AFB 8-3-1950 445 Hrs TT

46-274 27th FEG 5-28-1948 "J" McClellan AFB 4-28-1950 300 Hrs TT

46-275 27th FEG 6-11-1948 "J" Robbins AFB 7-31-1950 495 Hrs TT

46-276 27th FEG 6-18-1948 "J" Robbins AFB 7-31-1950 409 Hrs TT

46-277 27th FEG 6-18-1948 "J" Kelly AFB 1-6-1950 255 Hrs TT

46-278 27th FEG 7-28-1948 "J" Robbins AFB 8-1-1950 134 Hrs TT

46-279 27th FEG 6-11-1948 "J" Robbins AFB8-4-1950 541 Hrs TT

46-280 27th FEG 8-4-1948 Write-off Bergstrom AFB No Code 4-22-1949 230 Hrs TT

46-281 27th FEG 6-14-1948 "J" McClellan AFB 8-1-1950 412 Hrs TT

46-282 27th FEG 6-21-1948 "A" Bergstrom AFB 1-26-1950 413 Hrs TT

46-283 27th FEG 6-8-1948 "J" McClellan AFB 4-28-1950 533 Hrs TT

46-284 27th FEG 6-11-1948 "J" McClellan AFB 4-28-1950 491 Hrs TT

46-285 27th FEG 8-4-1948 "J" Robbins AFB 8-11-1950 483 Hrs TT

46-286 27th FEG 6-6-1948 "J" Robbins AFB 9-13-1950 455 Hrs TT

46-287 27th FEG 6-8-1948 "A" Bergstrom AFB 12-6-1949 357 Hrs TT

46-288 27th FEG 6-8-1948 "J" Robbins AFB 9-8-1950 423 Hrs TT

46-289 27th FEG 6-11-1948 Write-off Kearney AFB No Code 3-31-1949 9 Hrs TT

46-290 27th FEG 6-11-1948 "J" McClellan AFB 4-11-1950 421 Hrs TT

46-291 27th FEG 6-11-1948 "J" McClellan AFB 4-11-1950 189 Hrs TT

46-292 27th FEG 6-28-1948 "J" Robbins AFB 10-2-1950 617 Hrs TT

46-293 27th FEG 6-17-1948 "J" Robbins AFB 8-11-1950 478 Hrs TT

46-294 27th FEG 8-11-1948 "J" Robbins AFB 9-11-1950 538 Hrs TT

46-295 27th FEG 6-21-1948 "J" Robbins AFB 7-31-1950 457 Hrs TT

46-296 27th FEG 6-11-1948 "J" Robbins AFB 8-9-1950 626 Hrs TT

46-297 27th FEG 6-17-1948 "J" Robbins AFB 11-27-1950 548 Hrs TT

46-298 27th FEG 7-7-1948 "J" Robbins AFB 7-31-1950 433 Hrs TT

46-299 27th FEG 6-21-1948 "J" Robbins AFB 7-31-1950 579 Hrs TT

46-300 27th FEG 6-28-1948 "J" Robbins AFB 7-31-1950 432 Hrs TT

46-302 27th FEG 7-8-1948 Write-off Bergstrom AFB No Code 5-24-1949 186 Hrs TT

46-303 27th FEG 7-8-1948 "A" Bergstrom AFB 2-28-1950 455 Hrs TT

46-304 27th FEG 6-28-1948 "J" McClellan AFB 4-28-1950 459 Hrs TT

46-305 27th FEG 6-28-1948 "J" Robbins AFB 8-12-1950 309 Hrs TT

46-306 27th FEG 7-7-1948 "J" McClellan AFB 4-28-1950 443 Hrs TT

46-307 27th FEG 7-7-1948 "J" Robbins AFB 8-12-1950 398 Hrs TT

46-308 27th FEG 7-9-1948 Write-off Randolph AFB No Code 9-29-1948 No Hrs Listed

46-309 27th FEG 7-7-1948 Write-off Bergstrom AFB No Code 5-16-1949 225 Hrs TT

46-310 27th FEG 7-12-1948 "J" Robbins AFB 7-31-1950 441 Hrs TT

46-311 27th FEG 7-7-1948 "A" Bergstrom AFB 12-9-1949 297 Hrs TT

46-312 27th FEG 7-8-1948 "J" McClellan AFB 4-28-1950 442 Hrs TT

46-313 27th FEG 7-9-1948 "J" Robbins AFB 7-31-1950 420 Hrs TT

46-314 27th FEG 7-13-1948 "J" Robbins AFB 7-31-1950 406 Hrs TT

46-315 27th FEG 7-31-1948 "J" Robbins AFB 8-8-1950 517 Hrs TT

46-316 27th FEG 7-8-1948 "J" McClellan AFB 4-28-1950 459 Hrs TT

46-317 27th FEG 7-7-1948 "J" Robbins AFB 8-3-1950 577 Hrs TT

46-318 27th FEG 8-4-1948 "J" Robbins AFB 7-31-1950 441 Hrs TT

46-319 27th FEG 7-8-1948 "J" Robbins AFB 7-31-1950 518 Hrs TT

46-320 27th FEG 7-9-1948 "J" Robbins AFB 8-1-1950 486 Hrs TT

46-321 27th FEG 7-9-1948 "J" Robbins AFB 7-31-1950 303 Hrs TT

46-322 27th FEG 7-9-1948 "J" Robbins AFB 8-8-1950 429 Hrs TT

46-323 27th FEG 7-9-1948 "J" Robbins AFB 8-10-1950 561 Hrs TT

46-324 27th FEG 7-9-1948 "J" Robbins AFB 7-31-1950 462 Hrs TT

46-325 27th FEG 7-8-1948 "J" Robbins AFB 7-31-1950 456 Hrs TT

46-326 27th FEG 8-4-1948 "J" Robbins AFB 7-31-1950 503 Hrs TT

46-327 27th FEG 8-4-1948 "J" McClellan AFB 4-28-1950 436 Hrs TT

46-328 27th FEG 7-9-1948 Write-off Chanute AFB No Code 4-13-1948 125 Hrs TT

46-329 27th FEG 7-12-1948 "J" Robbins AFB 8-8-1950 370 Hrs TT

46-330 27th FEG 7-13-1948 "J" Robbins AFB 7-31-1950 484 Hrs TT

46-331 27th FEG 7-12-1948 "J" Robbins AFB 7-31-1950 550 Hrs TT

46-332 27th FEG 7-28-1948 "J" Robbins AFB 8-30-1950 552 Hrs TT

46-333 27th FEG 7-13-1948 "J" McClellan AFB 4-28-1950 439 Hrs TT

46-334 27th FEG 7-13-1948 "J" McClellan AFB 4-28-1950 455 Hrs TT

46-335 27th FEG 7-12-1948 "J" McClellan AFB7-31-1950 353 Hrs TT

46-336 27th FEG 7-13-1948 "J" McClellan AFB 7-31-1950 545 Hrs TT

46-337 27th FEG 7-13-1948 "J" Robbins AFB 8-9-1950 394 Hrs TT

46-338 27th FEG 7-13-1948 "J" Robbins AFB 8-9-1950 409 Hrs TT

46-339 27th FEG 7-13-1948 "J" Robbins AFB 7-31-1950 597 Hrs TT

46-340 27th FEG 7-28-1948 "J" McClellan AFB 4-28-1950 477 Hrs TT

46-341 27th FEG 7-28-1948 "J" McClellan AFB 4-28-1950 515 Hrs TT

46-342 27th FEG 7-13-1948 "J" Robbins AFB 9-8-1950 427 Hrs TT

46-343 27th FEG 8-4-1948 "J" Robbins AFB 8-3-1950 436 Hrs TT

46-344 27th FEG 8-4-1948 "J" Robbins AFB 9-7-1950 495 Hrs TT

46-345 27th FEG 8-4-1948 "J" Robbins AFB 8-8-1950 440 Hrs TT

46-346 27th FEG 7-28-48 "J" McClellan AFB 4-28-1950 427 Hrs TT

46-347 27th FEG 7-28-1948 "J" Robbins AFB 7-31-1950 455 Hrs TT

46-348 27th FEG 7-28-1948 "J" McClellan AFB 8-1-1950 304 Hrs TT

46-349 27th FEG 8-4-1948 "J" Robbins AFB 8-1-1950 475 Hrs TT

46-350 27th FEG 8-4-1948 "J" Robbins AFB 8-7-1950 506 Hrs TT

46-351 27th FEG 7-28-1948 "J" Robbins AFB 7-31-1950 496 Hrs TT

46-352 27th FEG 7-28-1948 "J" Robbins AFB 7-31-1950 473 Hrs TT

46-353 27th FEG 8-4-1948 Write-off Bergstrom AFB 4-26-1948 No Code 4-26-1949 229 Hrs TT

46-354 27th FEG 8-4-1948 "J" McClellan AFB 4-28-1950 494 Hrs TT

Note: The "J" codes listed for the 27th FEG are given as the dates that the aircraft were delivered to either Robbins AFB, Georgia of McClellan AFB, California USAF reclamation centers, instead of the actual scrapping dates. Final salvage of these aircraft had taken place by the end of April 1951.

52nd Fighter (All-Weather) Group

F-82 B

44-65177 6-28-1948 To Shepherd AFB (ATC) 6-1-1949

F-82F

46-407 9-8-1948 To 325th F(AW)G 11-6-1950

46-408 9-9-1948 To 325th F(AW)G 11-2-1950

46-409 9-10-1948 "J" Brooks AFB 12-12-1950 511 Hrs TT

46-410 7-1-1948 "J" Brooks AFB 4-11-1950 357 Hrs TT

46-411 9-9-1948 To 105th FIS Knoxville, TN 2-28-1951 To 39th ADW Elmendorf 4-15-1952

46-412 9-15-1948 "A" McGuire AFB 5-18-1950 248 Hrs TT

46-414 9-9-1948 "J" Brooks AFB 12-12-1950 525 Hrs TT

46-415 9-9-1948 To McClellan AFB for Conversion to F-82G standards To 68th F(AW)S

46-416 9-9-1948 "J" McGuire AFB 8-10-1950 260 Hrs TT

46-417 9-10-1948 "A" Mitchell AFB 8-22-1949 74 Hrs TT

46-422 9-9-1948 "A" Stewart AFB 8-20-1951 706 Hrs TT

46-424 9-10-1948 "A" Mitchell AFB 8-22-1949 161 Hrs TT

46-425 6-27-1950 From 325th F(AW)G "J" McGuire AFB 8-10-1950 320 Hrs TT

46-429 10-15-1948 To 325th F(AW)G 11-7-1950

46-433 8-8-1948 "J" McClellan AFB 12-28-1951 595 Hrs TT

46-435 10-1-1948 "J" Brooks AFB 4-11-1950 626 Hrs TT

46-438 6-27-1050 From 325th F(AW)G "J" McGuire AFB 8-10-1950 410 Hrs TT

46-440 10-1-1948 To Wright Patterson AFB 3-6-1951 as EF-82F

46-441 9-29-1948 To 325th F(AW)G 11-11-1950

46-442 9-29-1948 Write-off 4-26-1949 No Code 132 Hrs TT

46-443 9-29-1948 To 325th F(AW)G 11-7-1950 Returned to 105th FIS 12-27-1951 To 39th ADW Elmendorf 4-1-1952

46-445 9-28-1948 "J" Robbins AFB 4-18-1952 667 Hrs TT

52nd F(AW)G

46-446 10-14-1948 "J" Brooks AFB 4-11-1950 459 Hrs TT

46-448 6-22-1950 From 325th F(AW)G To McClellan AFB for conversion to F-82G standard. To 68th F(AW)S

46-452 10-1-1948 Write-off No Code McGuire AFB 9-14-1950 509 Hrs TT

46-454 9-28-1948 "A" Stewart AFB 5-18-1950 473 Hrs TT

46-457 5-21-1951 From 325th F(AW)G "A" McGhee-Tyson AFB 6-31-1951 713 Hrs TT

46-458 10-14-1948 "A" McGuire AFB 5-2-1950 364 Hrs TT

46-460 10-14-1948 "A" McGuire AFB 5-2-1950 151 Hrs TT

46-462 10-25-1948 To Olmstead AFB 9-25-1949 "E" Olmstead 12-13-1950 220 Hrs TT

46-466 10-25-1948 To 325th F(AW)G 11-2-1950 Returned to be assigned to 105th FIS 5-25-1951 To 39th ADW Elmendorf 4-1-1952

46-468 10-25-1948 Write-off 5-5-1949 No Code Mitchell AFB 143 Hrs TT

46-472 10-28-1948 "A" Stewart AFB 3-1-1951 655 Hrs TT

46-474 10-14-1948 Write-off 5-19-1949 No Code Mitchell AFB 60 Hrs TT

46-477 6-22-1950 From 325th F(AW)G "A" Stewart AFB 10-16-1951 648 Hours TT

46-479 10-25-1948 "J" Brooks AFB 4-21-1951 479 Hrs TT

46-486 6-20-1950 Fro, 325th F(AW)G To 325th F(AW)G 11-7-1950 Returned to 52nd F(AW)G 5-21-1951 To 105th FIS 11-1-1951 To 39th ADW Elmendorf 4-28-1952

46-487 10-25-1948 "J" Brooks AFB 4-21-1951 581 Hrs TT

Note There are four F-82F's shown as being assigned to the 105th Fighter Interceptor Squadron (46-411, -443, -466 and -486). The individual aircraft record cards also show that four additional F-82's were assigned to McGhee-Tyson AFB, TN, the location at which the 105th FIS was stationed during the Korean War. These were; 46-422, -445, -457 and -477. Officially, the 105th FIS only flew F-47's and F-51's during this time period. Did they actually fly the Twin Mustang as well? Further verification would be appreciated. The 105th FIS was attached to the 52nd Fighter Interceptor Wing, the parent unit for the 52nd F(AW)G, at the time of their activation into federal service.

325th Fighter (All-Weather) Group

F-82B

44-65173 from 27th FEG To ATC 5-3-1949

F-82F

46-407 11-7-1950 from 52nd F(AW)G "J" McChord AFB 3-9-1951 534 Hrs TT

46-408 11-2-1950 from 52nd F(AW)G "J" Brooks AFB 5-7-1951 735 Hrs TT

46-413 10-20-1948 No loss code or location 73 Hrs TT

46-418 10-28-1948 "J" Brooks AFB 8-5-1951 550 Hrs TT

46-419 9-23-1948 "J" Brooks AFB 7-8-1951 727 Hrs TT

46-420 9-23-1948 "J" McChord AFB 8-28-1951 740 Hrs TT

46-421 9-7-1948 To Chanute AFB for mock-up 4-15-1949 32 Hrs TT

46-423 9-2-1948 Write-off 10-1948 No Code 16 Hrs TT

46-425 9-23-1948 To 52nd F(AW)G 6-27-1950

46-426 9-3-1948 "A" Lowry AFB 10-12-1949 225 Hrs TT

46-427 9-23-1948 "A" Moses Lake AFB 1-13-1950 159 Hrs TT

46-429 11-7-1950 from 52nd F(AW)G "A" McChord AFB 12-8-1950 537 Hrs TT

46-430 9-23-1948 "A" Moses Lake AFB 7-10-1950 82 Hrs TT

46-431 9-23-1948 "J" Brooks AFB 7-6-1951 712 Hrs TT

46-432 9-23-1948 "A" Hill AFB 10-11-1951 369 Hrs TT

46-434 9-3-1948 "J" Hill AFB 12-23-1951 307 Hrs TT

46-436 10-15-1948 Write-off 3-1949 No Code 125 Hrs TT

46-437 9-3-1948 "J" McChord AFB 3-1-1951 689 Hrs TT

46-439 9-23-1948 Write-off McChord AFB 5-12-1949 No Code 53 Hrs TT

46-441 11-11-1950 From 52nd F(AW)G "J" McChord AFB 4-13-1951 694 Hrs TT

46-443 11-7-1950 From 52nd F(AW)G To 52nd F(AW)G 5-9-1951

46-444 10-18-1948 "J" Brooks AFB 5-7-1951 812 Hrs TT

46-449 10-15-1948 "A" McChord AFB 2-28-1950 29 Hrs TT

46-450 10-15-1948 "A" McChord AFB 6-22-1950 275 Hrs TT

46-451 10-15-1948 "J" Brooks AFB 4-25-1951 831 Hrs TT

46-455 10-15-1948 "J" Hill AFB 2-23-1951 571 Hrs TT

46-456 10-15-1948 "J" McChord AFB 8-10-1950 407 Hrs TT

46-457 10-15-1948 To 52nd F(AW)G 5-21-1951

46-461 10-15-1948 "A" McChord AFB 6-21-1951 659 Hrs TT

46-464 10-15-1948 "J" McChord AFB 6-22-1950 95 Hrs TT

46-466 11-2-1950 From 52nd F(AW)G To 52nd F(AW)G 5-25-1951

46-467 10-20-1948 "J" Brooks AFB 7-19-1951 793 Hrs TT

46-469 10-20-1948 "J" Brooks AFB 7-23-1951 703 Hrs TT

46-470 10-15-1948 "J" Brooks AFB 5-24-1951 632 Hrs TT

46-475 12-21-1948 "J" Brooks AFB 7-5-1951 763 Hrs TT

46-478 10-20-1948 "J" Brooks AFB 8-5-1951 862 Hrs TT

46-486 10-20-1948 To 52nd F(AW)G 6-2-1950 Returned 11-7-1950 To 52nd F(AW)G 5-21-1951 To 105th FS11-1-1951

319th Fighter (All-Weather) Squadron

The 319th F(AW)S F-82F's are listed separately since they were originally delivered to France Air Force Base, Canal Zone prior to the squadron joining the 325th F(AW)G at McChord Air Force Base.

(Date assigned France AFB/Date Assigned McChord AFB)
46-438 10-22-1948 4-24-1949 To 52nd F(AW)G 6-20-1950

46-488 10-22-1948 4-13-1949 To 52nd F(AW)G 6-22-1950

46-453 10-22-1948 4-13-1949 "J" Brooks AFB 7-5-1951 685 Hrs TT

46-459 10-22-1948 4-13-1949 Write-off 7-19-1950 No Code McChord AFB 375 Hrs TT

46-464 10-25-1948 4-29-1949 "J" Brooks AFB 7-24-1951 543 Hrs TT

46-471 10-25-1948 3-9-1949 "J" Brooks AFB 7-24-1951 472 Hrs

46-477 10-22-1948 5-12-1949 To 52nd F(AW)G 6-22-1950 325th F(AW)G

46-481 10-25-1948 4-13-1949 "A" McChord AFB 8-1-1950 435 Hrs TT

46-482 10-28-1948 4-15-1949 "J" Brooks AFB 5-7-1951 620 Hrs TT

46-483 10-22-1948 4-24-1949 "J" Brooks AFB 11-6-1950 429 Hrs TT

46-484 10-28-1948 4-13-1949 "J" Brooks AFB 7-16-1951 524 Hrs TT

46-485 10-22-1948 4-13-1949 "J" Brooks AFB 8-1-1950 251 Hrs TT

46-493 10-28-1948 4-13-1949 "J" Brooks AFB 8-1-1950 392 Hrs TT

46-494 10-28-1948 5-19-1949 "J" Brooks AFB 11-6-1950 348 Hrs TT

46-495 10-25-1948 4-13-1950 "J" Brooks AFB 8-1-1950 335 Hrs TT

4th, 68th and 339th Fighter (All-Weather) Squadrons (347th F(AW)G)

46-355 68th F(AW)S "M" 8-7-1950 (Moran/Meyer) 89 Hrs TT

46-356 4th F(AW)S "A" at Naha Air Base 8-10-1951 "Gluckin-Blackin." Ex 68th and 339th F(AW)S 654 Hrs TT

46-357 68th F(AW)S "P" 5-26-1951 (Nielson/Mulhallen)688 Hrs TT

46-358 4th F(AW)S "N" 6-30-1950 (Sayer/Lindvig) 122 Hrs TT

46-359 4th F(AW)S "A" at Naha Air Base 11-7-1950 (Tymonicz/Jennings) "Misguided Virgin" 311 Hrs TT

46-360 4th F(AW)S ? 9-28-1950 69 Hrs TT

46-361 68th F(AW)S Received from 4th F(AW)S 12-1950 Returned to 4th F(AW)S 4-1951 To 449th F(AW)S 5-1952

46-362 F(AW)S To 4th F(AW)S 6-1951 To 449th F(AW)S 6-1952

46-363 68th F(AW)S To 449th F(AW)S 5-1952 "Siamese Lady"

46-364 68th F(AW)S "M" 6-29-1950 Loss Code revised to "Q", Enemy action not on a combat mission, as aircraft was bombed at Suwon, South Korea. (Moran's) "B.O. Plenty" 37 Hrs TT

46-365 339th F(AW)S To 69th F(AW)S 6-1951 To 449th F(AW)S 6-1952

46-366 4th F(AW)S To 449th F(AW)S 5-1952

46-367 339th F(AW)S "A" at Johnson Air Base 3-2-1951 "Lover Boy" 325 Hrs TT

46-368 F(AW)S "A" at Johnson Air Base 5-7-1950 59 Hrs TT

46-369 68th F(AW)S "J" 5-11-1950 as Cannibalized for spare parts. 165 Hrs TT

46-370 13th Air Base Group "J"2-30-1950 as cannibalized for spare parts 9 Hrs TT

46-371 68th F(AW)S "N" 8-12-1951 at Kimpo, South Korea. Hit a bulldozer. "da Quake" 916 Hrs TT

46-372 339th F(AW)S To 68th F(AW)S 7-1951 To 449th F(AW)S 5-1952

46-372 339th F(AW)S To 68th F(AW)S 7-1951 To 449th F(AW)S 5-1952

46-373 68th F(AW)S "A" 2-12-1951 (Broughton) 67 Hrs TT

46-374 68th F(AW)S "A" 5-1-1951 at Itazuke Air Base 710 Hrs TT

46-375 68th F(AW)S "A" 12-7-1950 at P'ohang, South Korea (Jacobs) 362 Hrs TT

46-376 68th F(AW)S "A" 9-28-1950 at Itazuke Air Base Mid air collision with 16th FS F-80 (Stanton/McDonald) 274 Hrs TT

46-377 339th F(AW)S To 4th F(AW)S 6-1951 To 449th F(AW)S 5-1952

46-378 68th F(AW)S "M" 7-3-1951 (Johnson/Doll) 244 Hrs TT

46-379 339th F(AW)S To 68th F(AW)S 7-1951 To 449th F(AW)S 5-1952

46-380 339th F(AW)S "A" 5-30-1951 at Johnson Air Base 40 Hrs TT

46-381 339th F(AW)S "A" 7-21-1950 at Johnson Air Base 77 Hrs TT

46-382 4th F(AW)S To 449th F(AW)S 5-1952 "Night Takeoff"

46-383 68th F(AW)S To 449th F(AW)S 5-1952 First F-82 to pass 1,000 hour flying mark. Hudson's aircraft when scoring the first kill of the Korean War. "Bucket of Bolts"

46-384 4th F(AW)S To 449th F(AW)S 5-1952 "Wee Pea"

46-390 4th F(AW)S To 449th F(AW)S 5-1952 "Midnight Sinner"

46-391 68th F(AW)S "M" 9-10-1950 178 Hrs TT

46-392 339th F(AW)S To 68th F(AW)S 7-9-1951 "N" Suwon, South Korea 11-6-1951 (Buckhout) "Our Little Lass" 729 Hrs TT

46-393 68th F(AW)S To 449th F(AW)S 5-1952

46-394 4th F(AW)S to 68th F(AW)S 12-3-1950 "P" 3-4-1951 (Fluhr/Milhaupt) "Dottie Mae"542 Hrs TT

46-395 339th F(AW)S To 4th F(AW)S 6-1951 To 449th F(AW)S 6-1952

46-396 68th F(AW)S Write-off 3-31-1950 No Code 69 Hrs TT

46-397 4th F(AW)S To 68th F(AW)S 12-5-1950 To 4th F(AW)S 4-25-1951 To 449th F(AW)S 5-1952

46-398 339th F(AW)S "A" Johnson Air Base 12-31-1951 (Vahat) 201 Hrs TT

46-399 68th F(AW)S "P" 1-26-1951 (Ancil/Greer) ""L'il Bambi" 525 Hrs TT

46-400 4th F(AW)S To 68th F(AW)S 12-3-1950 "A" 12-7-1950 (Harding/Pratt) "Call Girl" 229 Hrs TT

46-401 339th F(AW)S To 68th F(AW)S 7-9-1951 To 449th F(AW)S 5-1952 "Gruesome Twosome" also "Patches on the Patches"

46-402 4th F(AW)S "M" 7-6-1950 Write-off as a result of battle damage 148 Hrs TT

46-403 4th F(AW)S To 449th F(AW)S 5-1952

46-404 339th F(AW)S To 68th F(AW)S 7-9-1951 To 449th F(AW)S 5-1952

The following F-82F's were modifies to F-82G standards and supplied to the 68th F(AW)S for action during the Korean War

46-415 68th F(AW)S From 52nd F(AW)G 5-10-1951 To 449th F(AW)S 5-1952

45-448 68th F(AW)S From 325th F(AW)G 7-27-1951 To 449th F(AW)S 5-1952

46-473 68th F(AW)S From Wright Patterson AFB 10-28-1951 To 449th F(AW)S 5-1952

46-491 68th F(AW)S From Eglin AFB 7-27-1951

To 449th F(AW)S 5-1952

449th Fighter (All-Weather) Squadron

F-82G

46-361 5-20-1952 From 4th F(AW)S "J" Elmendorf AFB 3-9-1953 704 Hrs TT

46-362 5-20-1952 From 4th F(AW)S "J" Elmendorf AFB 5-12-1953 605 Hrs TT

46-365 5-20-1952 From 4th F(AW)S "E" Ladd AFB 1-5-1953 669 Hrs TT

46-366 5-20-1952 From 4th F(AW)S "J" Elmendorf AFB 5-12-1953 673 Hrs TT

46-372 5-20-1952 From 68th F(AW)S "J" Elmendorf AFB 5-12-1953 755 Hrs TT

46-377 5-20-1952 From 4th F(AW)S "J" Elmendorf AFB 1-24-1953 697 Hrs TT

46-379 5-20-1952 From 68th F(AW)S "J" Elmendorf AFB 5-9-1953 705 Hrs TT

46-382 5-20-1952 From 4th F(AW)S "E" Ladd AFB 1-5-1953 651 Hrs TT

46-383 5-20-1952 From 68th F(AW)S "J" Elmendorf AFB 5-12-1953 1063 Hrs TT

F-82H

46-384 3-16-1949 From NAA "A" Ladd AFB 5-13-1951 369 Hrs TT

46-385 4-15-1949 "J" Ladd AFB 8-20-1952 456 Hrs TT

46-386 3-16-1949 "J" Ladd AFB 5-21-1953 848 Hrs TT

46-387 2-23-1949 "J" Ladd AFB 7-20-1953 438 Hrs TT

46-388 4-11-1949 "A" Ladd AFB 4-30-1950 159 Hrs TT

F-82G

46-389 5-20-1952 From 4th F(AW)S "J" Elmendorf AFB 5-12-1953 624 Hrs TT

46-390 5-20-1952 From 4th F(AW)S "J" Elmendorf AFB 11-2-1953 652 Hrs TT

46-393 5-20-1952 From 68th F(AW)S "E" Ladd AFB 1-5-1953 533 Hrs TT

46-395 5-20-1952 From 4th F(AW)S "E" Ladd AFB 1-5-1953 624 Hrs TT

46-397 5-20-1952 From 4th F(AW)S "J" Elmendorf AFB 5-19-1953 708 Hrs TT

46-401 5-20-1952 From 68th F(AW)S "J" Elmendorf AFB 5-8-1953 926 Hrs TT

46-403 5-20-1952 From 4th F(AW)S "J" Elmendorf AFB 5-9-1953 721 Hrs TT

46-404 5-20-1952 From 68th F(AW)S "J" Elmendorf AFB 5-9-1953 760 Hrs TT

F-82F

46-411 4-15-1952 From 105th FS "J" Elmendorf AFB 5-12-1953 718 Hrs TT

46-414 5-20-1952 From 68th F(AW)S "J" Elmendorf AFB 1-21-1953 893 Hrs TT

46-443 4-28-1952 From 105th FS "J" Elmendorf AFB 5-9-1953 855 Hrs TT

46-448 5-20-1952 From 68th F(AW)S "J" Elmendorf AFB 1-24-1953 916 Hrs TT

46-466 5-1-1952 From 105th FS "J" Elmendorf AFB 5-12-1953 655 Hrs TT

46-473 5-20-1952 From 68th F(AW)S "J" Elmendorf AFB 1-24-1953 328 Hrs TT

46-486 4-28-1952 From 105th FS "J" Elmendorf AFB 5-12-1953 648 Hrs TT

46-491 5-18-1952 From 68th F(AW)S "J" Elmendorf AFB 5-24-1953 364 Hrs TT

F-82H

46-496 3-16-1949 "J" Ladd AFB 8-20-1952 399 Hrs TT

46-497 2-28-1949 "A" Ladd AFB 1-19-1950 160 Hrs TT

46-498 4-12-1949 "E" Ladd AFB 7-3-1953 799 Hrs TT

46-499 4-11-1949 "E" Ladd AFB 8-27-1953 946 Hrs TT

46-500 3-16-1949 "J" Ladd AFB 4-8-1953 899 Hrs TT

46-501 4-29-1949 "J" Ladd AFB 8-20-1952 440 Hrs TT

46-502 4-15-1949 "E" Ladd AFB 5-24-1950 116 Hrs TT

46-503 5-9-1949 "E" Ladd AFB 6-8-1953 739 Hrs TT

46-504 4-11-1949 "J" Ladd AFB 8-20-1952 548 Hrs TT

APPENDIX D

Bibliography

449th All-Weather Fighter Squadron arrives at Ladd AFB Midnight Sun, April 1, 1949

C.M. Daniels, *The Twin Tailed Threat* Air Classics, Vol 5, No. 1. 14ff.

Warren Thompson, *North American Twin Mustang* Aeroplane Monthly, March 1978, p 116ff.

Warren Thompson, *F-82 Killers Over Korea* Air Enthusiast, Number 6, p 121 ff.

Warren Thompson, *Double Trouble* Wings, Vol. 13 No. 4 23 ff.

Norman L. Avery, *North American Aviation, Inc, Chronology of First Flights* Journal of the American Aviation Historical Society, Vol. 2 No. 2 p. 47 ff.

Garry R. Pape, *Escort Fighter* Airpower, Vol. 7 No. 6 P 48 ff.

Lin Hendrix, *Range Finder* Wings, Vol. 7 No. 2 p. 58 ff.

Robert Trimble, *Twin Mustang* Air Classics Vol. 10 No. 3 p 60 ff.

Robert Trimble, *Twin Mustang, Part II* Air Classics Vol. 10 No. 4 p 33 ff.

Jay Miller, *P-82 Restoration Nears Completion* Air Classics Vol. 12 No. 12 p 56 ff.

Col. John Sharp, *Flying the P-82 in Combat* Air Classics Vol. 10 No. 7 p 49 ff.

Lt. R.K. Walter, *Twin Mustang* Air Trails Vol. XVII No. 4 p 39 ff.

Boardman C. Reed, *The First Kill of the USAF* Journal of the American Aviation Historical Society Vol. 2 No. 2 p 72ff.

Stanley J. Grogan, Jr. *Lightning Lancers: Combat Highlights of the 68th Squadron in Korea* Aeroplane Historian 1961.

Merle C. Olmstead, *One Day in June* Journal of the American Aviation Historical Society Vol. 15 No. 1 p 42 ff.

Robert F. Futrell, *The United States Air Force In Korea* Duell, Slone and Pearce, New York 1961

M. J. Hardy, *The North American Mustang* Arco, New York 1979

William N. Hess, *Fighting Mustang the Chronicle of the P-51* Doubleday 1970.

Robert W. Gruenhagen, *Mustang: The Story of the P-51 Fighter* Arco, New York 1969

Robert Jackson, *Airwar Over Korea* Scribners, New York 1973

Maurer Maurer, *Air Force Combat Units of World War II* Zenger, Washington 1980

Combat Squadrons of the Air Force in World War II Zenger, Washington 1980

Handbook, Flight Operating Instructions, F-82F, F-82G, F-82H USAF, November 5, 1949. (Plus revisions through 1951)

Handbook, Flight Operating Instructions, F-82E, USAF, November 5, 1949

Standard Aircraft Characteristics. F-82E, USAF September 22, 1950

Standard Aircraft Characteristics. F-82F, USAF September 22, 1950

Standard Aircraft Characteristics. F-82G, USAF September 22, 1950

APPENDIX E

The F-82's Combat Performance

In spite of being encumbered by the radar pod, the F-82G had proven itself to be an effective tactical aircraft. The six .50 caliber machine guns in the wing's center section could lay down quite an accurate and withering fusillade. The aiming problem, engendered by the offset position of the cockpit, was simply compensated by adjustment of the K-18 gunsight. That and practice by the crews put the bullets where they were intended to go.

racks as was the case with the F-82E. This limited the G's to some extent. There were two problems discovered when the G's carried external loads. As has been noted earlier in the Moran incident, external tanks had a tendency to whip to the side and damage the ailerons instead of falling free. The other was simply a problem of pilot education. When firing rockets at night, they had to be fired individually instead of in a salvo or "ripple." Otherwise the crews would be blinded by the "rockets' red glare."

Described as a "pilot's airplane and a crew

Another combination of F-82 ordnance. Rocket "trees" on three stations, a chemical or smoke tank under the left wing, and a napalm tank under the other. (Note the fuse.) - NAA via Avery

Early on, there was a problem of machine gun barrels rupturing when the guns were fired. This created damage to the 82's wing and gun bay. It was solved by X-raying the barrels prior to installation. Combat ordnance loads depended on the mission but they normally consisted of the two outer wing racks carrying extra fuel while the inner bomb racks were fitted with "trees" that mounted five 5" HVAR rockets each. There was no problem with installing "trees" on the outer bomb racks if necessary. A bomb load of up to two thousand pounds could be carried or two napalm tanks containing 165 gallons each. For some unknown reason during the development of the F-82G, someone had decided that it wouldn't be necessary to carry bombs on the inner

chiefs' nightmare," the F-82 was stable throughout all flight attitudes. It **DID** require constant retrimming as fuel was consumed. The internal equipment, which was sophisticated for its day, always needed attention. The cooling system had a propensity for leaks and the Allison V-1710 engines always needed retuning. These were the same engines as used in the XP-51J and they featured a speed-density carburetor which required a delicate touch by the mechanic to keep them running. B.J. Buckhout states that he had only one engine failure in over eight hundred hours of pilot time in the F-82 – not a bad track record considering the amount of stress placed on the aircraft flying combat. The biggest problem with the F-82G/F was in the radar pod, which was pressur-

Attaching 2.75" aircraft rockets to a tree. The smaller rockets had to be slid onto the rail which was attached to the tree, while the larger rockets hooked directly onto the posts. Note the mounting of the 100 pound bomb to the curved porion of the outer rack, which was actually curved to the contour of the normal 165 gallon tank. - USAF via Greenhalgh

Shown here is an experimental rocket rack. The bottom rocket had to be fired first, and if it didn't fire, then the upper rocket would not ignite, supposedly! Sometimes they did anyway, which resulted in some wildly erratic missile flight in front of the aircraft. The rocket racks were retractable, but the program was dropped for obvious reasons. - NAA via Avery

ized with nitrogen. A leak would render the radar's magnetron useless which meant that the aircraft was out of service as an all-weather fighter. All three of the 347th's squadrons had these problems and all felt that they were "doing good" if they could get a third of their aircraft serviceable due to pod failures. The difficulty could have been averted by pressurizing the pod off of an available "pad" on the Allison engine but the USAF would not permit the modification.

his eye. This was a particularly strong while watching the opposite fuselage "come around you" in a roll.

Instrument flight was smooth except for a couple of "non sequiturs" in design or USAF logic. The pilot's cockpit held one instrument that was critical to instrument flight – and all pilots feared and hated it. It was the J-8 Attitude Gyro and it read exactly opposite to the presentation of all other conventional gyros. When the miniature airplane on the instru-

A pair of rocket launcher "trees" on an F-82B. Rockets in this case 5" HVAR's, were fired in a predetermined sequence from the inner racks first, followed by the outer racks. They could not be jettisoned , but had to be fired off in case of an emergency.
-Avery

The F-82G was exceedingly manueverable, with power-boosted controls. It was flown just like any other twin-engined aircraft with a few exceptions. The flight instruments were calibrated to those in the left, or pilot's, fuselage. If the pilot flew with reference to his instruments, he could have a "good ride." This was fine for him but bad for his radar observer. A thoughtless pilot could whip his R/O into submission quite readily if he didn't off-set the "needle and ball" one needle's width, particularly during rolls. Keeping the needle centered in the pilot's cockpit would result in a sensation of being slung around in a centrifuge to the hapless R/O. Then too, there was always the impending feeling of one about to have a mid-air collision when the pilot caught sight of the right fuselage out of the corner of

ment was **BELOW** the reference line, the aircraft was actually **CLIMBING!** During an instrument approach, the pilot was forced to continually reorient his thinking as to what his Twin Mustang was actually doing. Bad news!

Another weirdo on the pilot's instrument panel was the Consolan Receiver (or the Visual Audio Range instrument). It was a low frequency version of today's VOR. It was positioned smack-dab in the position where '82 aircrews would have preferred the F-1 autopilot to be installed. In their infinite wisdom, the USAF wouldn't approve the repositioning of the autopilot. What makes this decision even harder to understand – **THE ONLY VAR TRANSMITTER IN THE WORLD WAS LOCATED IN OHIO!**

	XP-82	XP-82A	XP-82B	P-82C	P-82D	P-82E	P-82F	P-82G	P-82H
Engines									
Merlin V-1650-23/25	X		X	X	X				
Allison V-1710-119		X							
Allison V-1710-143/145						X	X	X	X
Propeller									
Aeroproducts									
A-524F-D1 (Left)	X								
AL-524-D1 (Right)	X								
Undetermined		X	X	X		X	X	X	X
Supercharger									
Integral - 2 stage -2 speed			X	X	X				
Integral engine stage single speed & auxiliary variable speed						X	X	X	X
Surface Control Boost									
Elevator and rudder only			X	X	X				
Elevator, rudder and aileron						X	X	X	X
Fire extinguisher						X	X	X	X
Anti Ice									
Provision only			X	X	X				
Wing, Empennage & Windshield						X	X	X	X
Propeller De-Ice									
Electric (Under boots)						X	X	X	X
Bomb Salvo									
Mechanical				X					
Electric						X	X	X	X
Radar									
SCR-720				X				X	X
APS-4					X				
AN/APG-28							X		
Tail Warning									
APS-13				X	X		X	X	X
Autopilot									
APN-1		A-10		X	X				
F-1							X	X	X
Cockpit									
Right cockpit equipped for emergency control						X			
Controls and instruments for radar operation only				X	X		X	X	X
Radar Altimeter									
AN//APN-1							X	X	X
Gunsight									
K-18		K-14				X	X	X	X
IFF									
SCR-695B						X			

Specifications

Model	Fuel Cap Gallons	Aux. Fuel	Wingspan (Feet)	Wing Area (Sq. Ft.)	Length (Feet)	Max Speed @ 21,500 Ft.	Cruise Speed	Initial Climb (Ft./Min.)	Service Ceiling	Empty Weight.	Normal Weight.	Maximum Weight.
XP-82	576 USG	Provision	51.6	408	42.2	482	227		41,600		19,100	
XP-82A	576 USG	Provision	51.3	408	42.2	468	227	4900	40,500	13,402	19,170	
P-82B	576 USG	Provision	51.6	408	42.2	482	227		41,600		19,100	
P-82C	576 USG	Provision	51.6	408	42.2	482	227		41,600		19,100	
P-82D	576 USG	Provision	51.6	408	42.2	482	227		41,600		19,100	
P-82E	576 USG	To 620 (2)*	51.2	417.6	39.1	404	264	4020	38,400	14,914	20,741	24,813
P-82F	576 USG	To 620 (2)*	51.6	417.6	42.2	400	248	3690	38,500	16,309	22,080	26,708
P-82G	576 USG	To 620 (2)*	51.6	417.6	42.2	400	248	3770	39,900	15,977	21,760	25,591
P-82H	576 USG	To 620 (2)*	51.6	417.6	42.2	400	250	3740	37,100	16,147	22,060	26,041

* Indicates 2 tanks

Index

About the Author

David R, McLaren grew up listening to "Terry and the Pirates," "Hop Harrigan" and best of all, "Captain Midnight" on the radio. And what better inspiration could there have been for a career in aviation? Born and raised in Springfield, Illinois, McLaren then spent thirty years as an air traffic controller in the USAF and the FAA in the mid-west. During those years, ha also raised Kevin. Eric and Christa along with attending Waubonsee College and Aurora University, where he majored in English and History. Currently retired from what he considered a most interesting occupation, he now believes that writing aviation history is his true vocation. He pursues this endeavor from his boyhood home in Springfield where he resides with his wife Sharon.

About the Artist

Gerald Asher is a third generation aviator; his grandfather owned a number of private aircraft between the 1930's and 40's. His father was a corporate pilot for over three decades. A self-taught artist, Gerald is a participant in the U.S. Air Force Art Program and a charter member of the American Society of Aviation Artists in which he currently serves as Chairman of Public Relations. He spends much of his spare time involved in flying and restoring World War II vintage aircraft. Born and raised in Michigan as the eldest of seven children, he now claims Fort Worth, Texas as home where he resides with his wife, Meg and their sons Ronald and Philip.

COVER PAINTING

"Ole '97", the aircraft assigned to the C.O. of the 27th Fighter Escort Group, is seen in a left break for runway 13 at Bergstrom AFB, Texas on a winter afternoon sometime in 1949.